20 14 Edition

End User License Agreement

This book (the "Book") is a product provided by MechboxPRO.com o/b AirsoftPRESS (being referred to as "MechboxPRO" in this document), subject to your compliance with the terms and conditions set forth below. PLEASE READ THIS DOCUMENT CAREFULLY BEFORE ACCESSING OR USING THE BOOK. BY ACCESSING OR USING THE BOOK, YOU AGREE TO BE BOUND BY THE TERMS AND CONDITIONS SET FORTH BELOW. IF YOU DO NOT WISH TO BE BOUND BY THESE TERMS AND CONDITIONS, YOU MAY NOT ACCESS OR USE THE BOOK. MECHBOXPRO.COM MAY MODIFY THIS AGREEMENT AT ANY TIME, AND SUCH MODIFICATIONS SHALL BE EFFECTIVE IMMEDIATELY UPON POSTING OF THE MODIFIED AGREEMENT ON THE CORPORATE SITE OF MECHBOXPRO.COM. YOU AGREE TO REVIEW THE AGREEMENT PERIODICALLY TO BE AWARE OF SUCH MODIFICATIONS AND YOUR CONTINUED ACCESS OR USE OF THE BOOK SHALL BE DEEMED YOUR CONCLUSIVE ACCEPTANCE OF THE MODIFIED AGREEMENT.

Restrictions on Alteration
You may not modify the Book or create any derivative work of the Book or its accompanying documentation. Derivative works include but are not limited to translations.

Restrictions on Copying
You may not copy any part of the Book unless formal written authorization is obtained from us.

LIMITATION OF LIABILITY
MechboxPRO will not be held liable for any advice or suggestions given in this book. If the reader wants to follow a suggestion, it is at his or her own discretion. Suggestions are only offered to help.

IN NO EVENT WILL MECHBOXPRO BE LIABLE FOR (I) ANY INCIDENTAL, CONSEQUENTIAL, OR INDIRECT DAMAGES (INCLUDING, BUT NOT LIMITED TO, DAMAGES FOR LOSS OF PROFITS, BUSINESS INTERRUPTION, LOSS OF PROGRAMS OR INFORMATION, AND THE LIKE) ARISING OUT OF THE USE OF OR INABILITY TO USE THE BOOK. EVEN IF MECHBOXPRO OR ITS AUTHORIZED REPRESENTATIVES HAVE BEEN ADVISED OF THE POSSIBILITY OF SUCH DAMAGES, OR (II) ANY CLAIM ATTRIBUTABLE TO ERRORS, OMISSIONS, OR OTHER INACCURACIES IN THE BOOK.

You agree to indemnify, defend and hold harmless MechboxPRO, its officers, directors, employees, agents, licensors, suppliers and any third party information providers to the Book from and against all losses, expenses, damages and costs, including reasonable attorneys' fees, resulting from any violation of this Agreement (including negligent or wrongful conduct) by you or any other person using the Book.

Miscellaneous.

This Agreement shall all be governed and construed in accordance with the laws of Hong Kong applicable to agreements made and to be performed in Hong Kong. You agree that any legal action or proceeding between MechboxPRO and you for any purpose concerning this Agreement or the parties' obligations hereunder shall be brought exclusively in a court of competent jurisdiction sitting in Hong Kong.

Preface

AIrsoft has been well established in Asia for over 20 years, but only in recent years has there been much interest in the Western world. Most professional literature on Airsoft were written in Japanese, with very few translated works available. Even though we are seeing more and more tech tips (in English) popping up on the internet these days, too many of them were written by newcomers who don't really know what they are talking about. If you upgrade your AEG based on their advices, you may risk running into unexpected troubles.

At MechboxPRO (a subsidiary of AirsoftPRESS), we produce technical information based on input from practicing engineers, technicians and field operators who have been with Airsoft since the era of S.S. 9000. Because we are part of the industry, we know what information is really needed, and we make sure our books tell what people really need to know. We do not mind to criticize thing that doesn't work, and we will not hesitate to give you hacks and workarounds to difficult problems. Reading this book should be like having an airsoft professional by your side, passing on useful hints whenever you get stuck.

About this Ultra FPS & ROF book

This Ultra FPS & ROF series book is an extension to the Advanced Mechbox Guide marketed under the AirsoftPRESS brand name. With feedback from customers and reviewers we further develop the original mechbox guide to additionally cover manufacturer specific tech info for FPS/ROF optimization.

To achieve ultra high FPS, you need to make sure the entire mechbox mechanism is optimized in such a way that the desired FPS can be produced without sacrificing ROF and battery life. High FPS performance must be achieved with practicality in mind – you should not need to trade off power with other performance factors, and the gears should not get striped too easily.

On the other hand, to achieve ultra high ROF you'll need to make sure the entire mechbox mechanism is optimized in such a way that piston movement is fast enough to produce the rate of fire desired. That means you need to use techniques to reduce the friction introduced due to fast movement, to ensure the spring can rebound fast enough after compression, and to reduce the overall workload of the gears and the motor.

Are you ready for the challenge? ☺

Are you ready for the upgrade works?

Upgrade works can cost you $$$$$. Some upgrade works can also put you into legal troubles. Therefore, before doing anything on the mechbox you better first assess your readiness on several key issues. Start by considering the issue on legitimacy. Is it legal to buy and own airsoft in your area? For teenagers: are your parents going to chew you up for an "unapproved" airsoft purchase?

Legitimacy	
Do they allow airsoft at all?	This is the most important question to ask if your sole source of income is your parents.
Is it legal to own and play airsoft in your area?	Some regions do not allow the purchase of airsoft. Some regions do not allow the possession of airsoft at all. Will you be sent to jail due to airsofting? The best thing to do is to call the local police department and ask. For your peace of mind, ask for a written confirmation from a police officer.
Power limit	If airsoft is allowed, what is the power limit? Check your local laws to make sure you are doing legal upgrades.
Restrictions on full auto shooting	Some countries do not alow full auto airsoft. New Zealand is an example. Some impose power restriction specifically on AEGs. Again, check your local laws.

NOTE: ## Case study: New Zealand

Air-powered weapons (airsoft guns are air powered) are legal to possess and use in New Zealand, provided that the person is either over 18 years of age, or 16 with a firearms license. A person under 18 may not possess an air gun but may use one under the direct supervision of someone over 18 or a firearms license holder.

It is illegal to use these weapons in any manner that may endanger or intimidate members of the public (pointing, brandishing, etc) except where there is reasonable cause, such as an Airsoft game.

<Police, New Zealand, Airguns Factsheet, http://www.police.govt.nz/service/firearms/info sheet04.html >

The next question to ask is, what are you going to use your airsoft gun for?

"Application"	
What are you going to use the AEG for?	If it is for self-defense or for security (i.e. as an alternative to using real steel), you want a

	seriously upgraded gun that is reliable. If it is for real serious target practicing, you want a longer rifle which can deliver the sort of accuracy you need. If it is for backyard fun only then anything will do ☺ Long VERSUS Short rifle would make a strategic difference in your upgrade goal. Longer rifles place focus primarily on FPS and accuracy, while shorter rifles may opt for higher ROF. We will talk about this later.
If you are going to skirmish frequently with your friends, what do your friends own?	If all your friends use springer only, there is no point of owning a serious AEG. If you guys are forming a team, ask yourself what role you are going to take. A sniper does not use a shorty after all!

NOTE:  When **MAJOR** FPS improvement is the focus, you may want to use 0.25g BBs. 0.20g BBs are way too light at this power level and will not deliver the accuracy you need. If you plan to go over 450FPS+, consider the use of 0.30g BBs for maximum flight path stability. Marushin has some very nice 0.30g BBs (USD$18 something per bag of 1800... ☹).

By the way, heavy BBs do drag down FPS quite a bit (due to the heavier weight) but may produce higher impact (also due to the heavier weight).

> They are VERY expensive though... you should really take this cost element into consideration when defining your desired upgrade level.

Closely linked to the "application" issue is budget. How much $ are you going to spend? Remember, if an AEG costs you $100, you should have at least $300 handy since you will need to buy upgrade parts and tools. You will also need spare $ to deal with parts breakage and all other sorts of expenses.

Budget	
How much $ do you have for the initial investment?	You need enough $ to buy an AEG, a set of battery and charger, and loads of ammos. And you need to have enough $ to get them shipped to your dwelling.
Do you have regular cash inflow?	This is about TCO (Total Cost of Ownership). Some AEGs require more maintenance than the others. You need $ to buy parts, and you need $ to buy tools. Upgraded guns are more powerful and are always more troublesome maintenance-wise! <u>To illustrate:</u> *Our recommendation for maintenance of UPGRADED AEGs (at a minimum) is that for every 20000~30000 rounds, replace the following items:* ● *the hopup rubber and the hopup spacer*

(bucking)
- *the motor brushes*
- *the spring*
- *the piston head (the O ring in particular)*
- *the bushings(if they have cracks or they no longer sit on the shell firmly).*
- *the air nozzle (if its tip starts to deform)*

Our recommendation for stock AEGs (at a minimum) is that for every 30000 rounds, replace the following items even if they aren't broken:

- *the spring*
- *the piston head (the O ring in particular)*
- *the bushings(do this primarily on poor quality metal bushings or plastic bushings)*

* also replace the motor brushes and the hopup rubber after about 40000~50000 rounds.

Upgradeability

In the world of V7 architecture, TM guns are the preferred choice upgradeability-wise. They are precisely built so you would seldom come across clueless surprises along the upgrade process. Put it this way – you put in a stiffer spring and you will see improved performance. Pretty straight forward. This is not like those cheap China-made clones – you put in an upgrade part which is supposed to work and it turns out to not work for no reason. There are small precision issues (plenty of them indeed) with the cheaper clones that could make things un-workable.

Sadly, $ is everything. With a limited budget you may be forced to forget about big names like TM, CA and ICS. Just keep this in mind - buying a cheap AEG would force you to spend $$$ on endless trouble fixing in the long run. a TM is like a Toyota, while a cheap clone is like a Daewoo. You can tell the difference.

Below you will find some general advices on the sort of basic "tune up" activities that are deem necessary for most stock V7 packages.

At the absolute minimum, what should be done once your stock package has arrived to ensure proper working order?	
TM	Nothing special has to be done on a new-in-the-box TM gun.

	It is basically maintenance free for the first 15000~20000 rds. Avoid using full size 9.6V battery. Mini 9.6V or full size 8.4V is fine.
China-clone	Quality varies quite significantly batch by batch. Don't go over 9.6V under stock configuration.

If all that you want is to replace the stock spring with something slightly more powerful, check out the table below. We are a little conservative when compiling this table since zero trouble is the goal here:

What initial spring upgrade can be done on these guns without the need for changing other parts (not even the stock bushings)?	
TM	A PDI 130% or a Guarder SP90 spring upgrade
China-clone	A PDI 130% or a Guarder SP90 spring upgrade

Practical Considerations

Some people suggest that you buy a MP5K if you are mostly doing CQB, or a M4 if you are going to shoot a lot in the woodland. We strongly suggest that you don't base your decision on this. The reasons are:

1. You don't always play pure-CQB or pure-woodland – you will likely do both.

2. If you are gaming at the stock power level, the difference in power and range between a MP5K and a M4 wouldn't be significant.

3. You can always make your AEG "longer". For example, you can install a longer inner barrel into a MP5K and hide it with a mock suppressor. On the other hand, it would be quite difficult for you to cut a M4 short.

4. We honestly believe that a user's body height should be given serious consideration when choosing his weapon. A short guy with a long rifle may lead to poor mobility. A tall guy with a shorty may look extremely weird...

Ease of maintenance is a major factor to consider. Talking about disassembly/reassembly, nothing beats the MP5s and the G3s. **M14 is very complicated to disassemble. Parts compatibility for V7 is less satisfactory.**

NOTE:  Magazine capacity is another factor that deserves your attention. Think about it, a 200rd MP5 mag costs almost the same as a 460 rds G36 mag. If your gun shoots reasonably fast, you may empty a 200rds mag in 5 seconds or less. How many mags are you gonna need? And how much $ would you need to spare for these mags?

Apart from the capacity issue, you need to know that longer mags tend to jam more easily. The longer the path the bullets need to travel before entering the chamber, the more likely it is for them to encounter debris (such as dirt particles) that blocks their way up.

Warning: Why should you care about the length of the rifle? If you have a long barrel, you will need a large-volume cylinder as a bigger mass of air is required to push the bullet out of the barrel. The load is heavier for the entire system. Pushing for high ROF under such condition would be technically possible but highly costly.

Buying pre-upgraded AEG VS D.I.Y.

This is kind of an "outsourcing" decision. The reality is that there isn't really much you can do to quickly health-check a pre-upgraded AEG. Problems usually arise after about 3000 shots. The best thing you can do is to ask for a 14-day guarantee so you can take the gun home and test it out thoroughly. If the gun does not give you any trouble in the first 5000 rds, chance is that it will be trouble-free for another 10000~15000 shots at the least. Keep in mind that heavily upgraded guns usually have shorter life spans. Going over 350fps starts to make things stressful, while 400fps+ means killing the gun bit by bit (UNLESS you are willing to spend shit load of $ to upgrade almost everything inside and outside of the mechbox).

Pre-upgraded or D.I.Y?	
Pre-upgraded	Someone with experience (hopefully) did the job for you. However, you may not be able to tell what have been installed into your AEG. Some shops also do not have staffs that are technically competent enough to do the upgrade (they put the customers at risk).
D.I.Y.	Valuable learning process. You can choose exactly what to put into the gun. However, improper upgrade can damage the parts and lead to big time frustration.

The proper ways of taking things apart and putting them back together

To avoid getting into chaotic situation when doing your assembly and disassembly works, try to be as organized as possible. Do your work on a clean, dry and flat surface which is close enough to reach without having to walk back and forth to access the necessary tools.

In fact, it is best to work things out on a big white towel. The towel provides a color contrast, thus making it easy to see the parts as you lay them down.

Before you remove each part, ask yourself the following questions and take notes if possible:

1. What is this, and what is it for?
2. Why is it made the way it is?
3. How tightly is it screwed on out-of-the-box? How hard is it to remove?

As you remove each part, lay it down on a clean flat surface in clockwise order, with each part pointing in the direction it laid when it was in place. Assign each part a number indicating the order in which it was removed. When you are ready to put them back together, start with the last part you removed and then go counterclockwise through the rest of the parts. If possible, go to your manufacturer's website (or any other web site) and look for any documentation files (mostly in PDF format) they offer for free download. Many of them include very detailed fly-out diagrams, a complete list of all parts as well as where they fit. This will greatly lessen any confusion you might have when putting things back together.

Warning:

!

Along the process of disassembling / assembling your mechbox, there are chances for unexpected debris to fly out (e.g. when you try to put a M120 into the mechbox, your spring guide may accidentally fly out with strong impact). Therefore:

- order your children to stay away from the work area.
- don't work in a location too close to the windows (you don't wanna break the window glass).
- wear safety goggles yourself if you are going to deal with very stiff springs.

Good luck!

Getting the tools you're going to need

The two basic types of screwdrivers are standard (slot / flat head) screwdrivers and Philips screwdrivers. Make sure you have drivers of different size handy – a whole set of drivers from your local store should cost less than USD$5 each. Only TM mechboxes use Torx screws. Torx head size is typically described using the capital letter "T" followed by a number, such as T5, T10, T15 and T25. TM mechboxes use T10 screws. You therefore need to have the corresponding driver handy.

Note: It doesn't hurt for you to use the regular Philips screws in place of the Torx screws. The Torx screws have nothing special nor unique other than the special screw head layout.

 NOTE: **Is it OK to use flat head screw driver to drive the mechbox torx screws?**

Why not? In fact, as long as the size fits, a flat head driver will work just fine on the torx screws. Do make sure you don't use brute force when driving the screws.

Screws with a centre hole that is hexagonal require the use of Allen wrenches or Hex drivers. To the best of our knowledge, however, no TM

---

mechbox shall require the use of hex screw.

You do need to perform motor positioning via a hex screw (on the TM model in particular) as shown below.

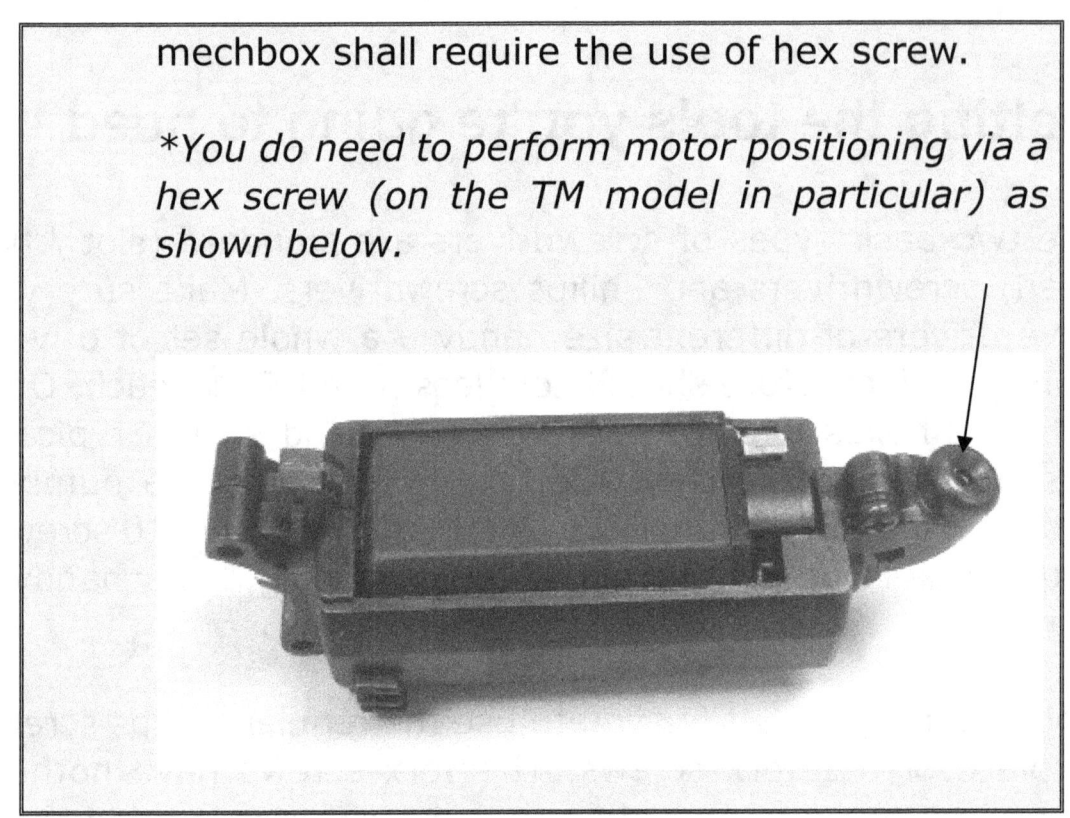

Proper handling of the mechbox screws

Most amateurs tend to screw things very tightly in hopes that the parts will not fly off later. This is in fact a deadly mistake because some screws, bolts and nuts are NOT supposed to be tightened too securely or the threads would be stripped.

If you find yourself confronted with a screw that is extremely difficult to get unscrewed, don't use brute force (or you risk stripping the threads). Instead, give the screw a slight twist in the opposite direction and then loose it again. If this does not help, tap the screwdriver on the head with a small hammer (but don't tap it too hard). If it still fails to make it, try to squirt the screw with penetrating oil or WD-40 and retry.

Stripping screws that go into places like the connection points between the mechbox and the pistol grip can be frustrating. If unfortunately you strip a screw, the 2 easy ways to remove it are:

- Fill the stripped screw hole with J.B. Weld (which is a type of glue specially for use with metal parts) or similar kind of product, and then put your screwdriver into the old hole to create a new fitting. Give it 10 to15 minutes to set and dry completely, then unscrew it.

- Drill a hole in the screw, then scoop out the old screw.

On the other hand, to prevent certain critical screws from getting loose, apply threadlock/locktite - a glue type of compound that makes screws more secure. Loctite usually won't bite into plastic very well. It can sometimes soften the plastic, but most of the time It won't really be permanently stuck on there. Most of the time you can get a locked screw loose via the use of a decent screwdriver (by the way, heat is what is used to release excessively strong locktite).

 NOTE: *Anywhere screws thread into metal, apply a dab of locktite for preventing all the vibration from loosening the screws.*

There are locktites of different strength available. You may want to check out the proper type to use through this URL: http://www.loctiteproducts.com/glue.asp

Other tools
You shall need needle-nosed pliers when handling the smaller mechbox screws and springs.

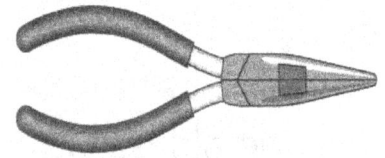

You may need some assorted files (such as straight edge, rounded and rounded side) for deburring and cleaning the edges of cuts needed for slight modification of certain mechbox parts.

You may need to use a pen shape soldering iron for rewiring the mechbox and the battery connection (especially when you are not happy with the existing wiring or you want to switch to the Deans plugs).

Desoldering is the process of removal of solder. It requires the application of heat to the solder joint and removing the molten solder so that the joint may be separated. You basically need to apply the desoldering iron tip onto the joint, then wait for the solder to liquefy, and finally have the molten solder removed through the pump.

If you want to use 7mm or larger ball bearings on standard TM shells, you must enlarge the space holders a bit through an electric dremel tool (those hand held type electric drill like tools for home use would just be fine).

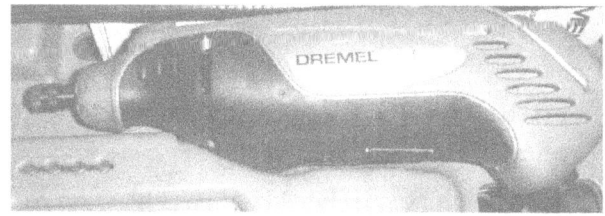

Soldering techniques

The goal of soldering is to join electrical parts together for forming an electrical connection. This is done via the use of a molten mixture of lead and tin (solder) with a soldering iron. Basic supplies needed for proper soldering include a soldering iron (the prong of metal that heats to a specific temperature through electricity), the soldering wire (an alloy of aluminum and lead), and a cleaning resin called flux that ensures the joining pieces are incredibly clean (by removing all the oxides on the surface of the metal that would interfere with the molecular bonding, allowing the solder to flow into the joint smoothly). A perfectly soldered joint should be nice and shiny looking, and should be very reliable in service. The key factors affecting the quality of the joint primarily include cleanliness, temperature, and adequate solder coverage.

The first step in soldering is cleaning the surfaces (including the iron tip itself). They must all be clean and free from contamination. Then, you may melt flux onto the parts to be joined. The parts should both be heated above the melting point of the solder but below their own melting point with the soldering iron. When touched to the joint, this precise heating can cause the solder to flow to the place of highest temperature and makes a chemical bond. Do keep in mind that too much solder is an unnecessary waste but too little may be insufficient for supporting the component properly.

You will find solder paste very useful along the soldering process:

AirsoftPRESS (Hong Kong).

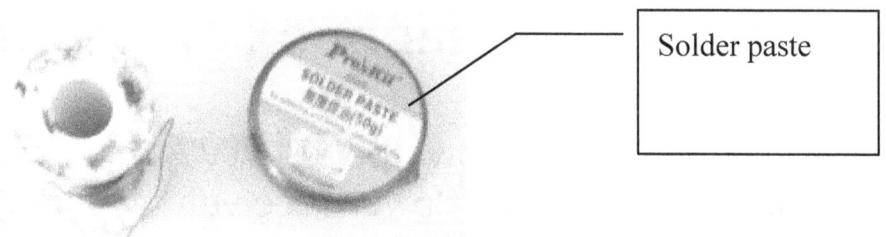

Solder paste

A solder paste is comprised of powdered metal solder suspended in a thick medium, which is known as flux. This flux is added to act as a temporary adhesive for holding the components together until the soldering process is finished.

Solder melts at around 190 degrees Centigrade. Such a temperature is hot enough to inflict a nasty burn. Be extremely careful when you do your soldering work. Also, do your soldering work in a room with good air circulation. Soldering does release toxic fumes.

If you are soldering battery connections, you may want to use a soldering pen of a high WATT value. The one shown below says 60W, which is powerful enough for most soldering jobs. Based on our experience, 20W is way too weak and 40W is only marginally sufficient. A minimum of 60W is recommended.

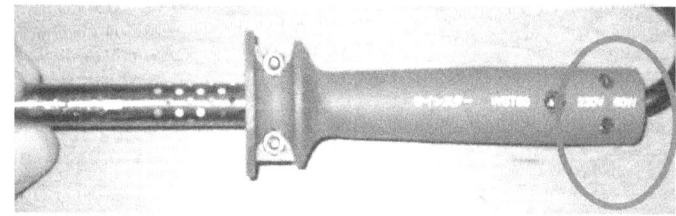

Overview of the M14 mechbox structure

These pictures show exactly how the M14 V7 mechbox is "attached" to the upper receiver (lower receiver removed).

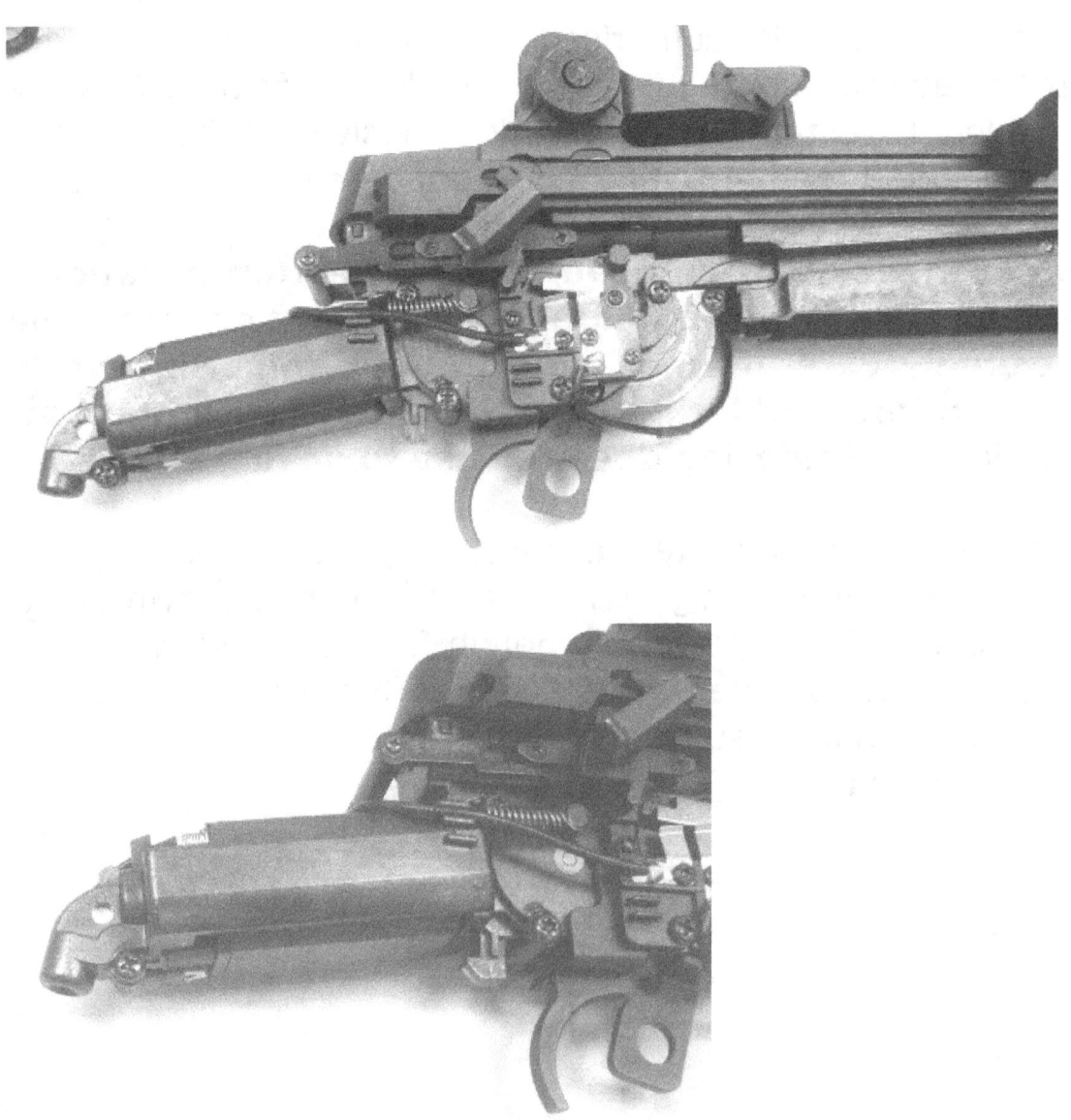

Overview of the select fire architecture

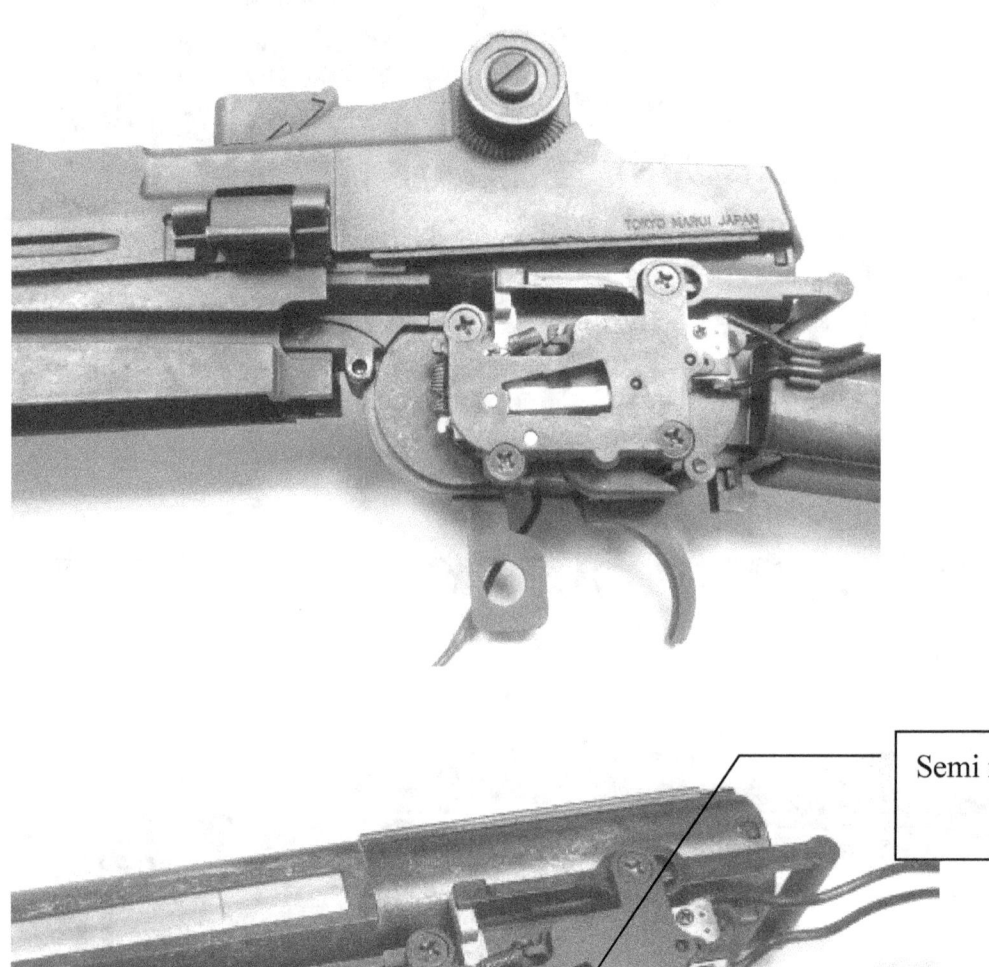

Semi mode

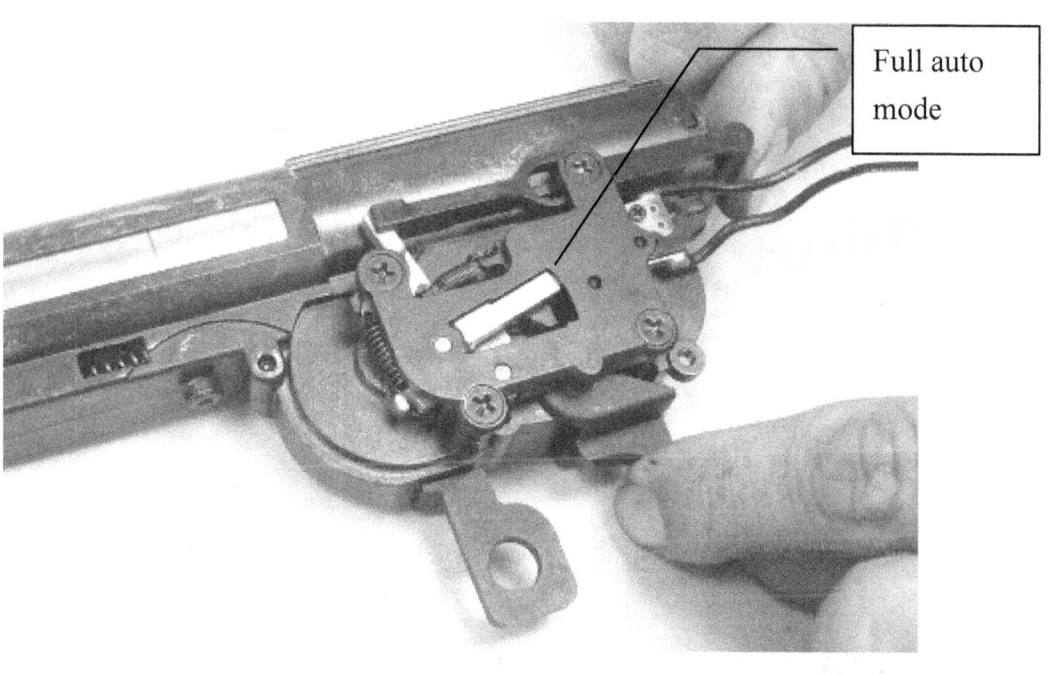

Full auto
mode

Note how these contact plates interact.

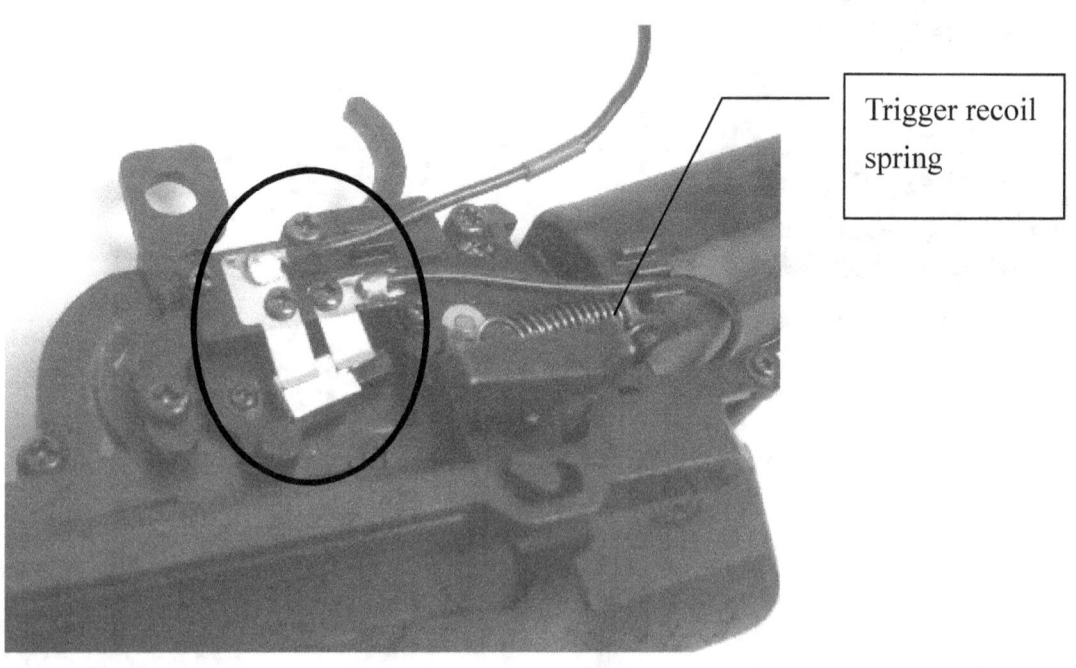

Trigger recoil spring

Step 0: Mechbox disassembly

Removing the motor mount

The pictures below are pretty self explanatory:

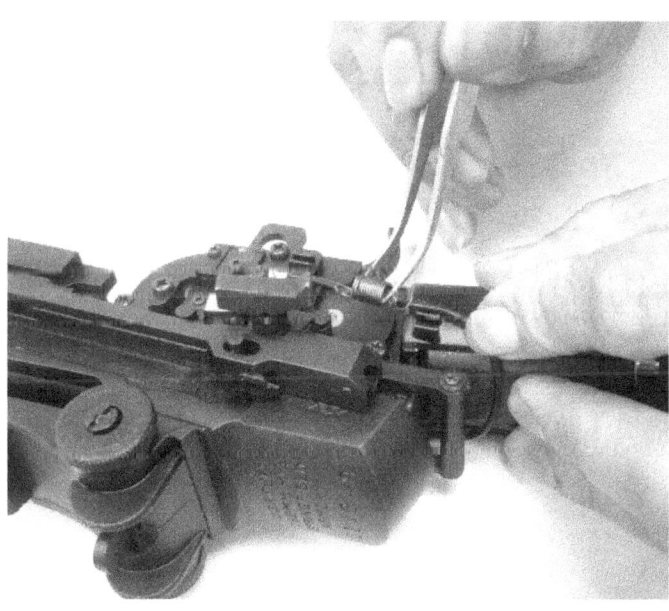

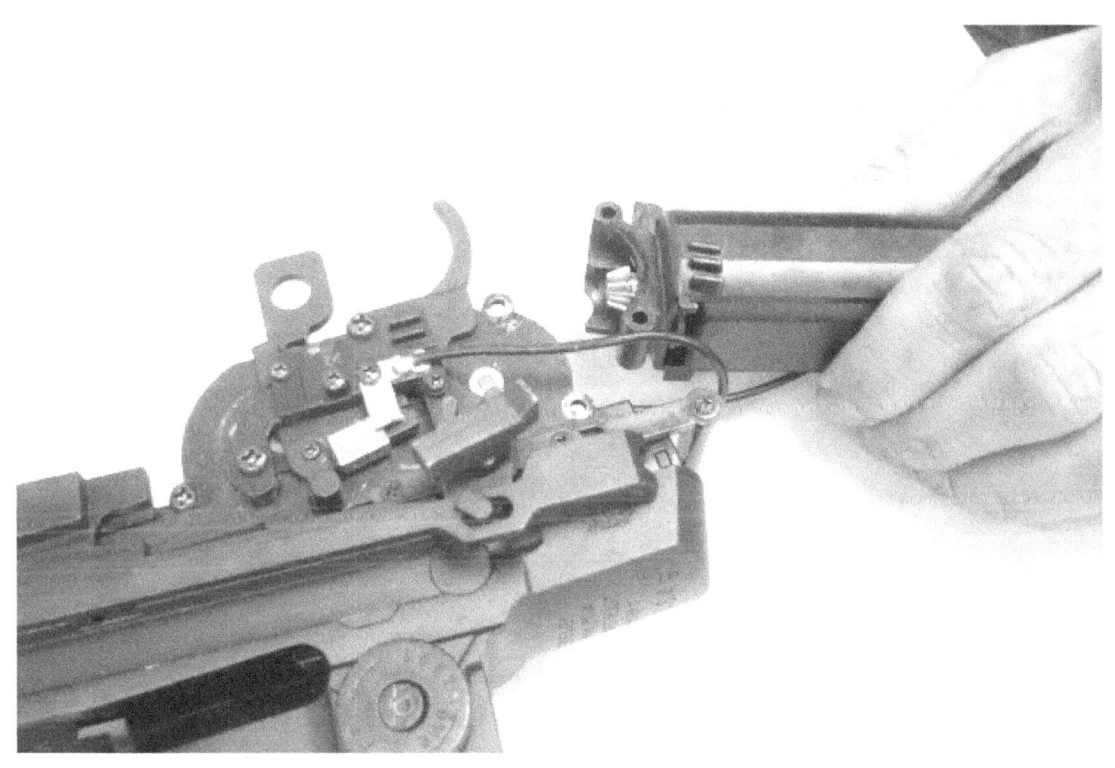

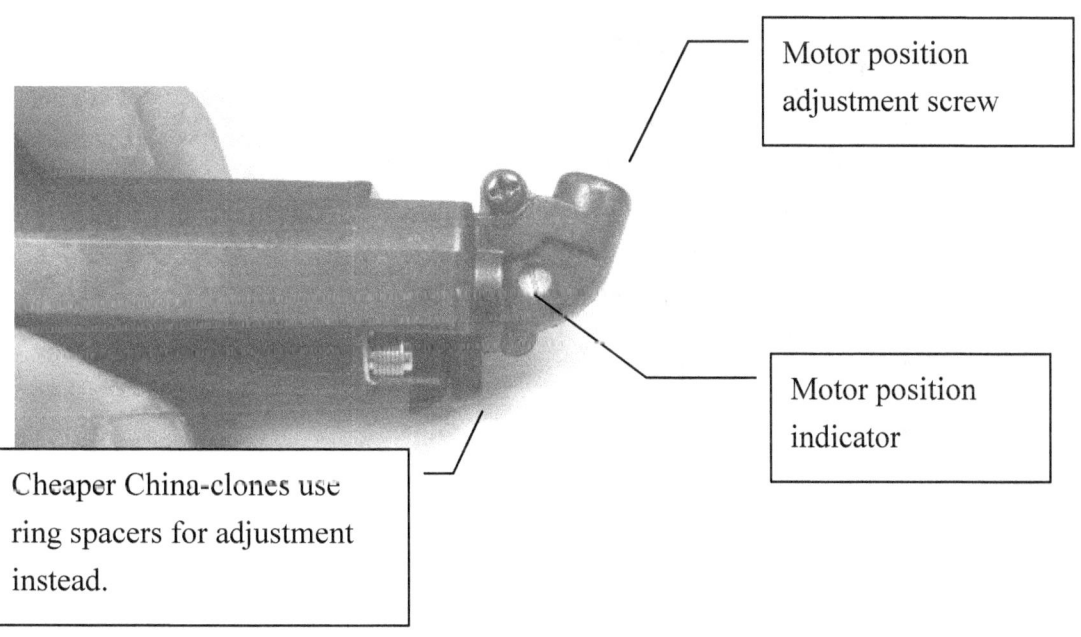

Motor position
adjustment screw

Motor position
indicator

Cheaper China-clones use
ring spacers for adjustment
instead.

The bevel gear is now exposed.

Detaching the upper receiver assembly

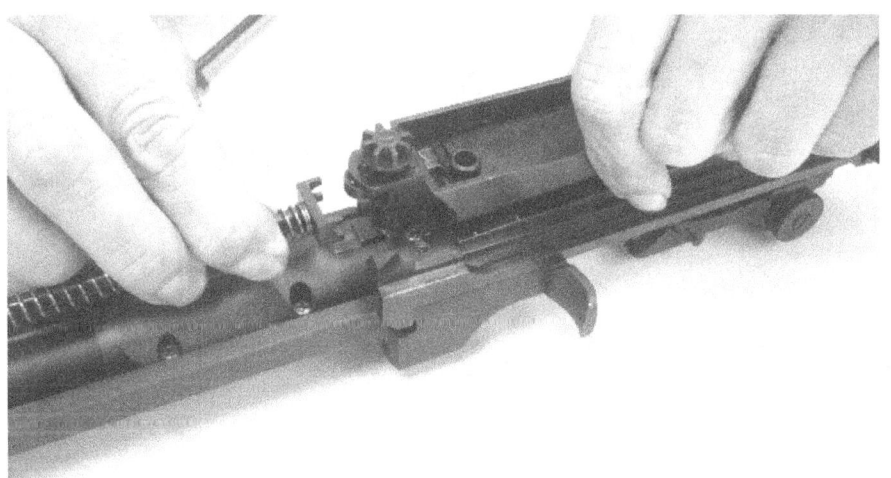

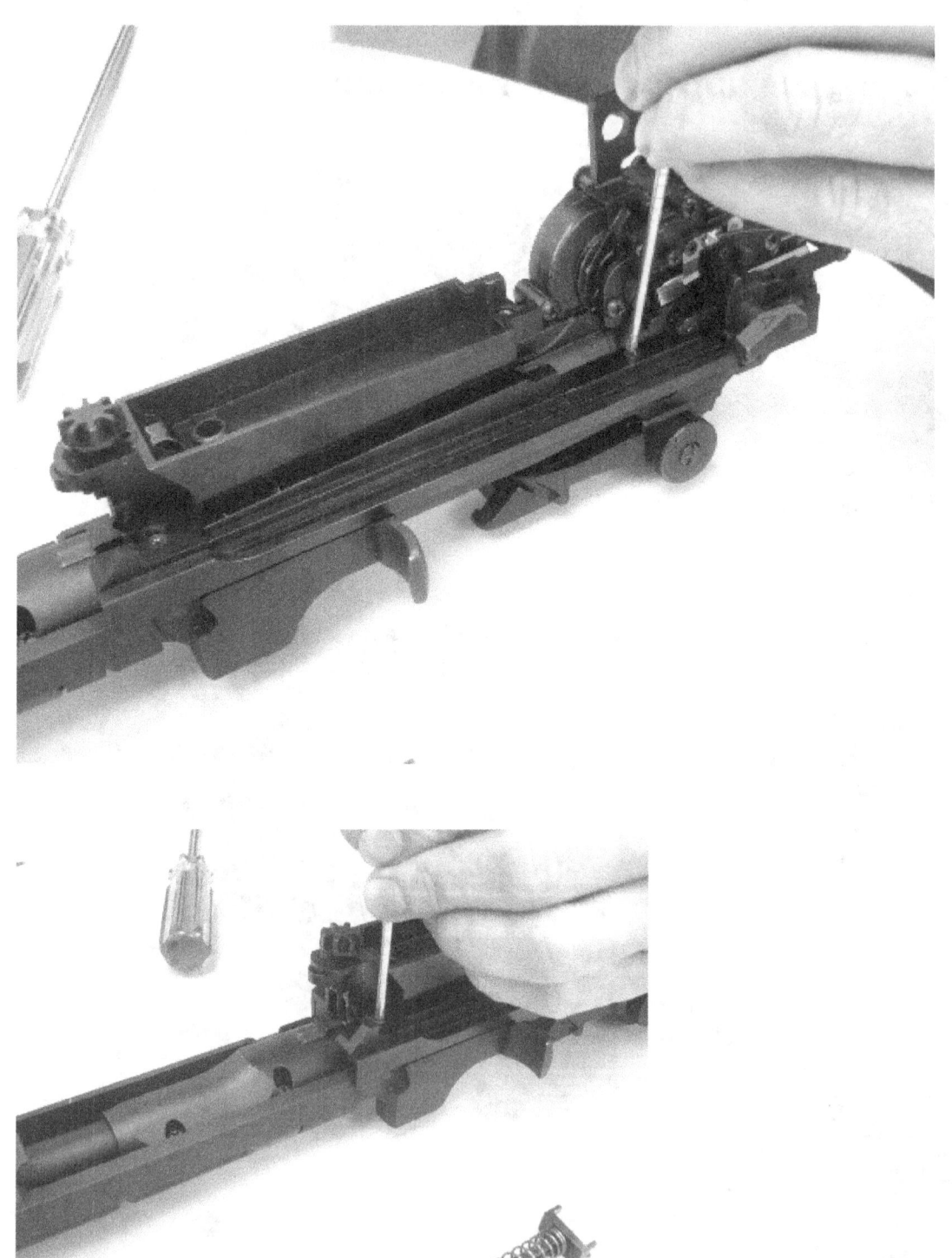

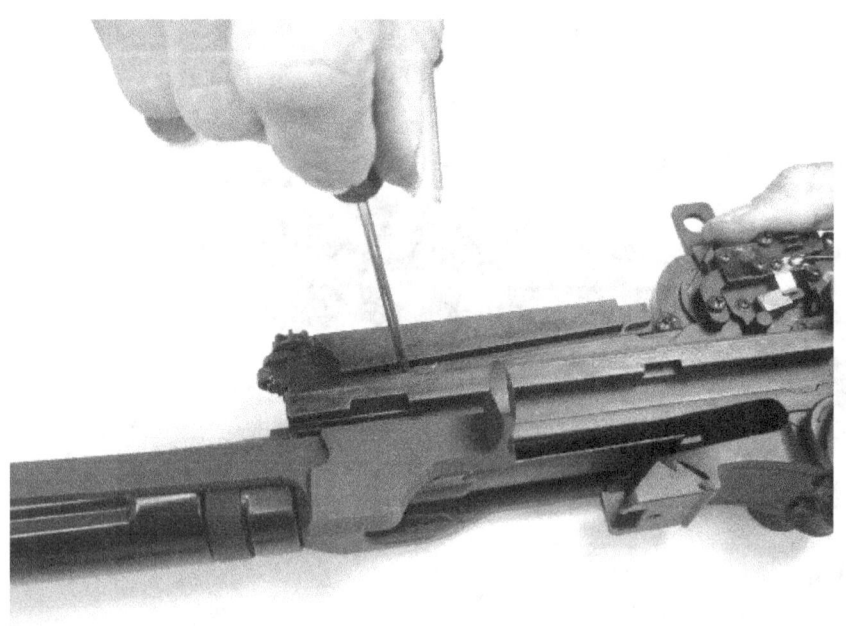

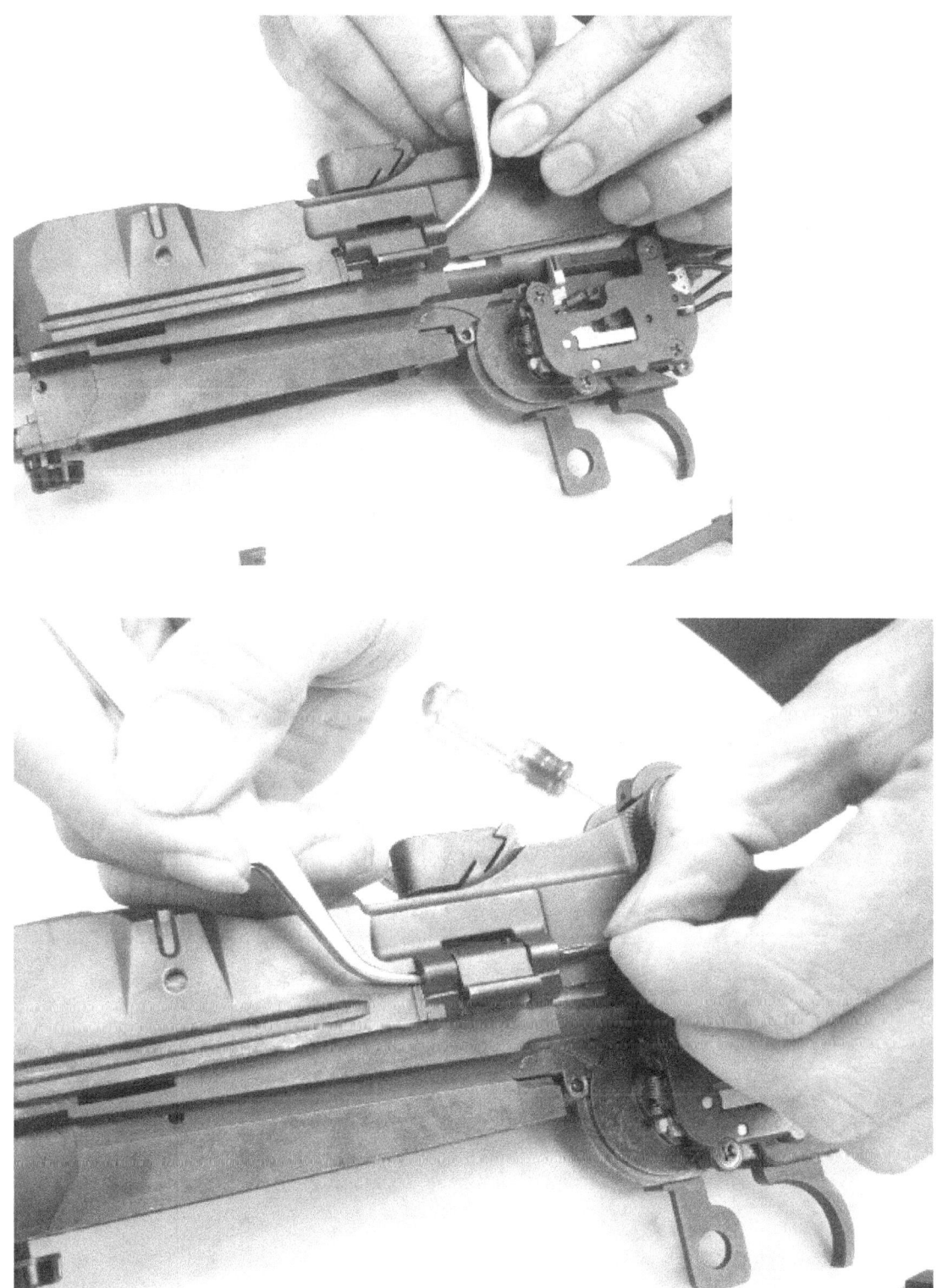

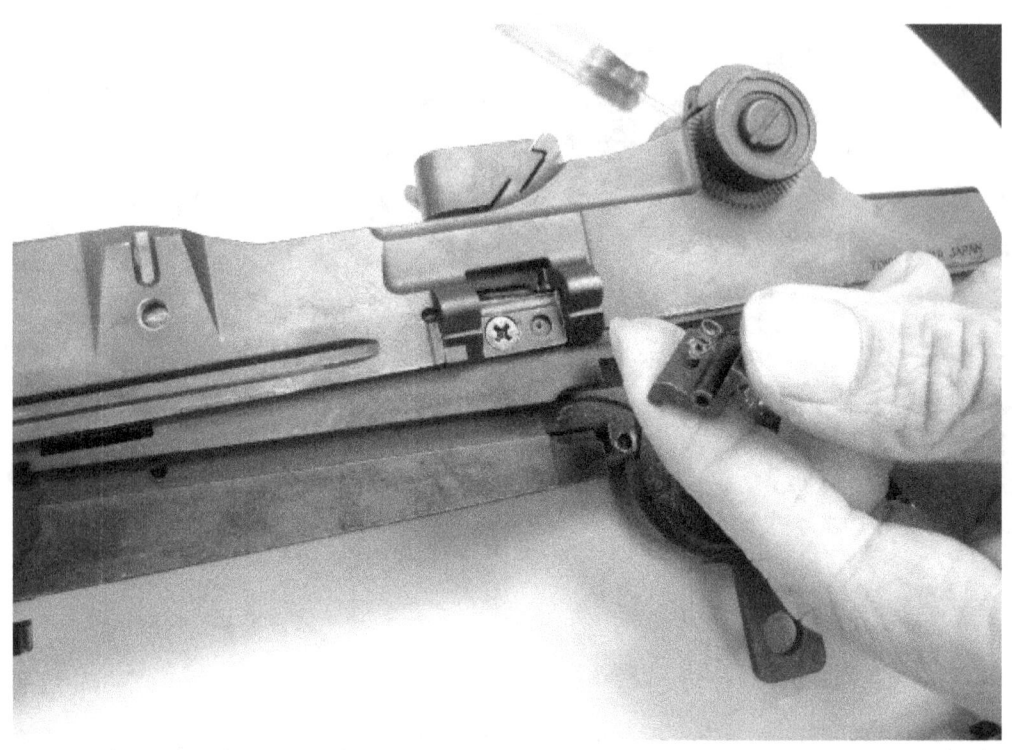

Removing the switch unit

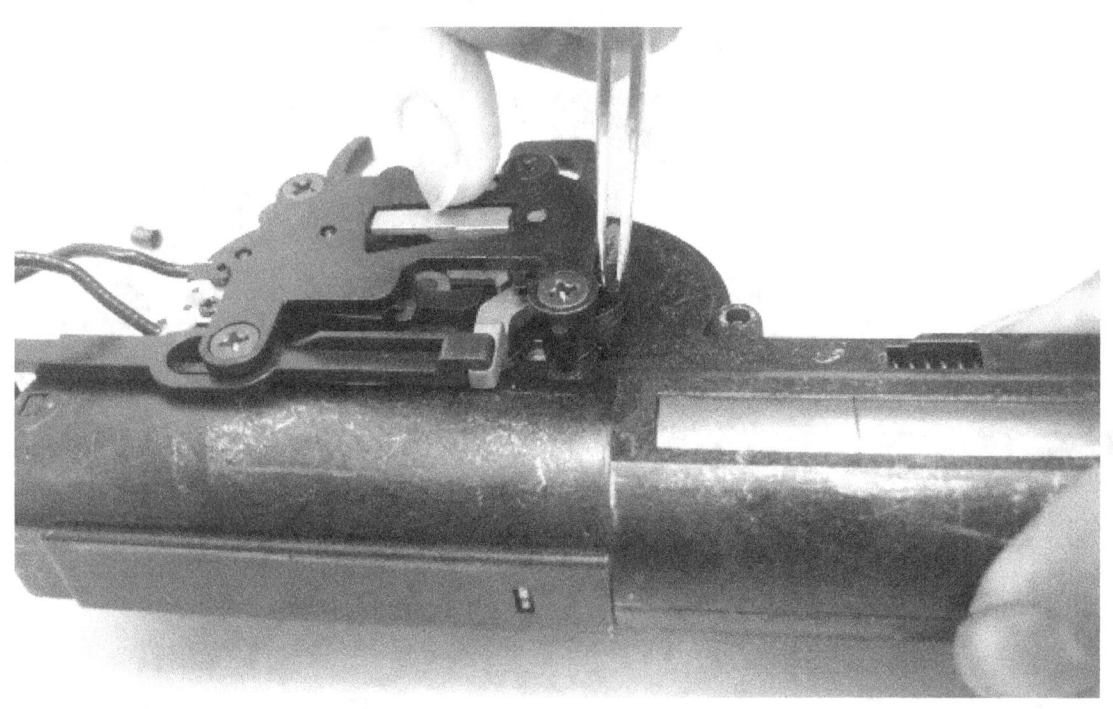

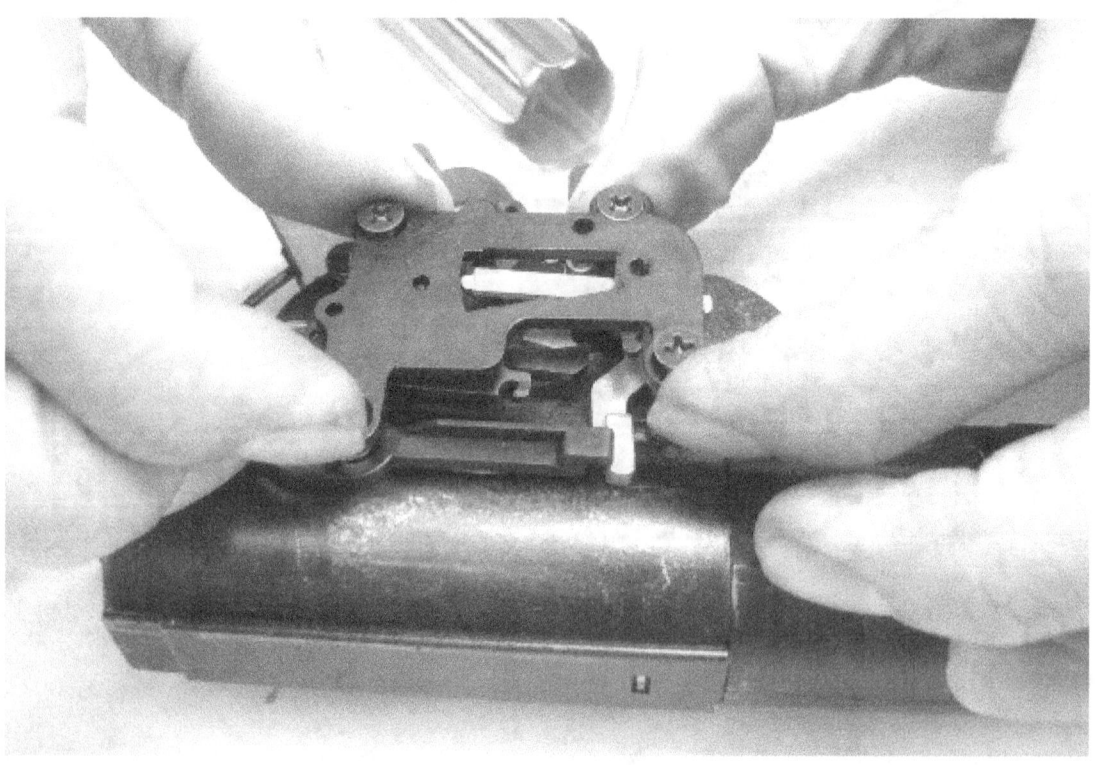

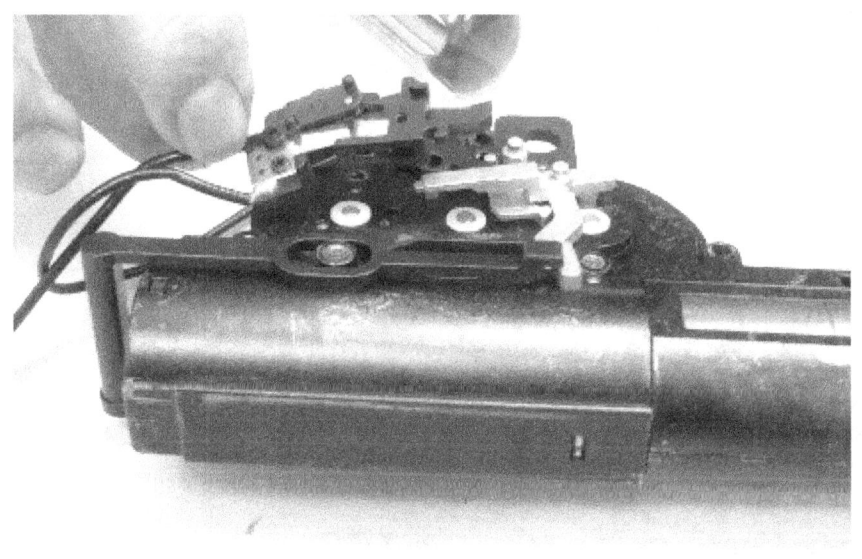

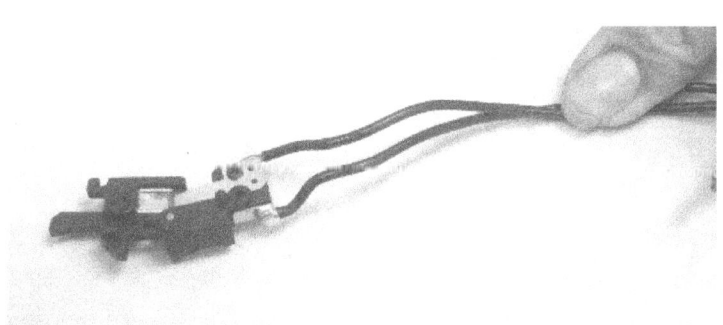

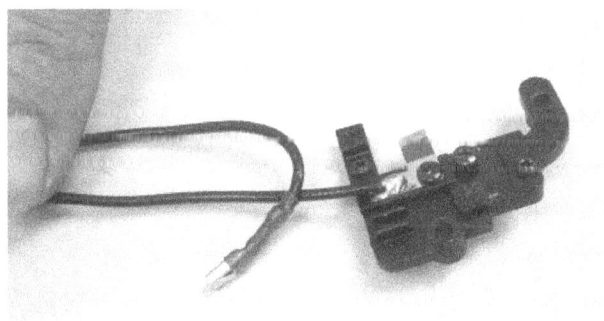

Removing the trigger assembly

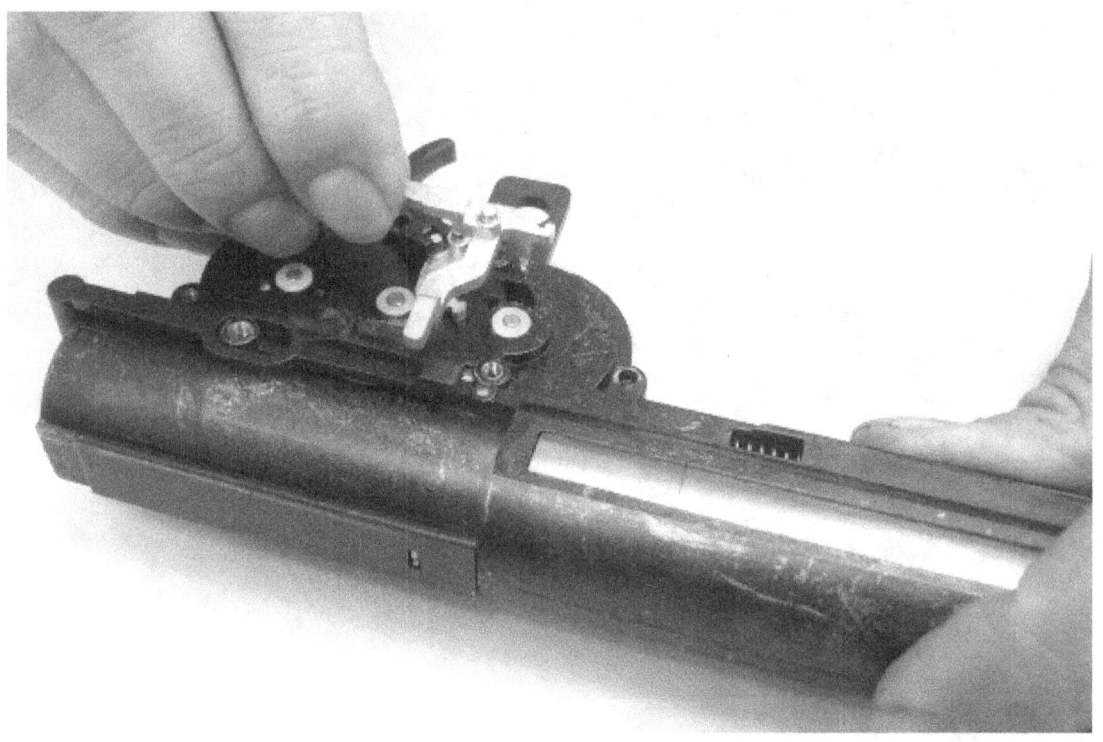

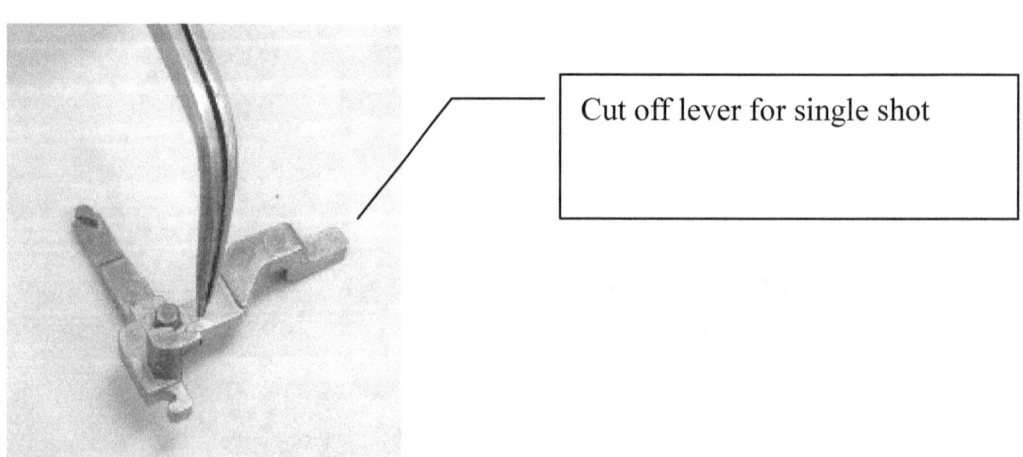

Cut off lever for single shot

The trigger unit

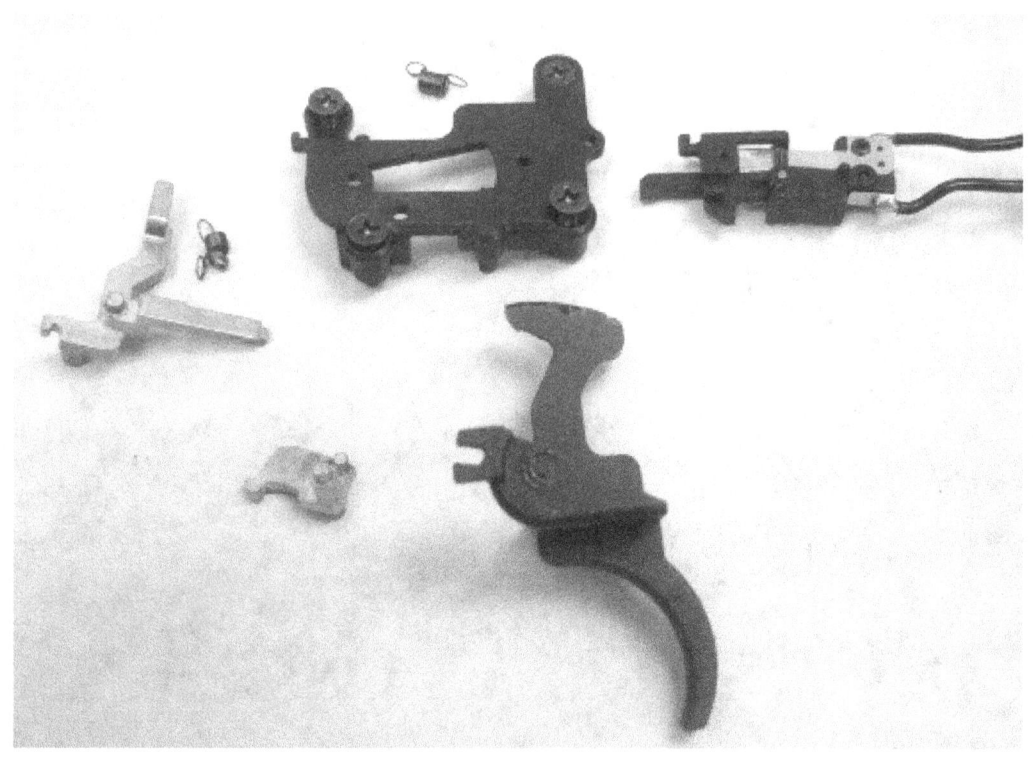

Opening up the shell

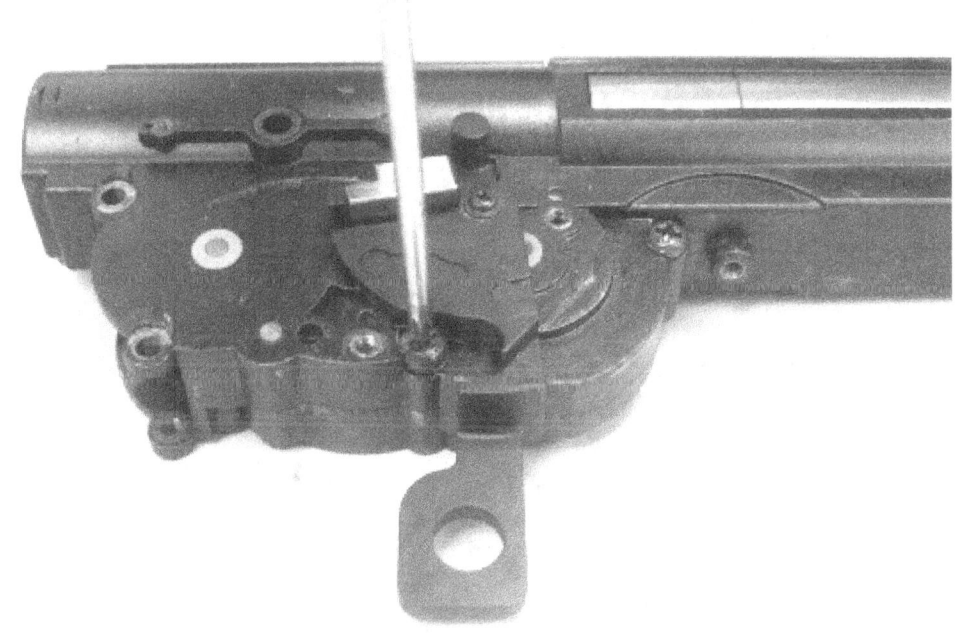

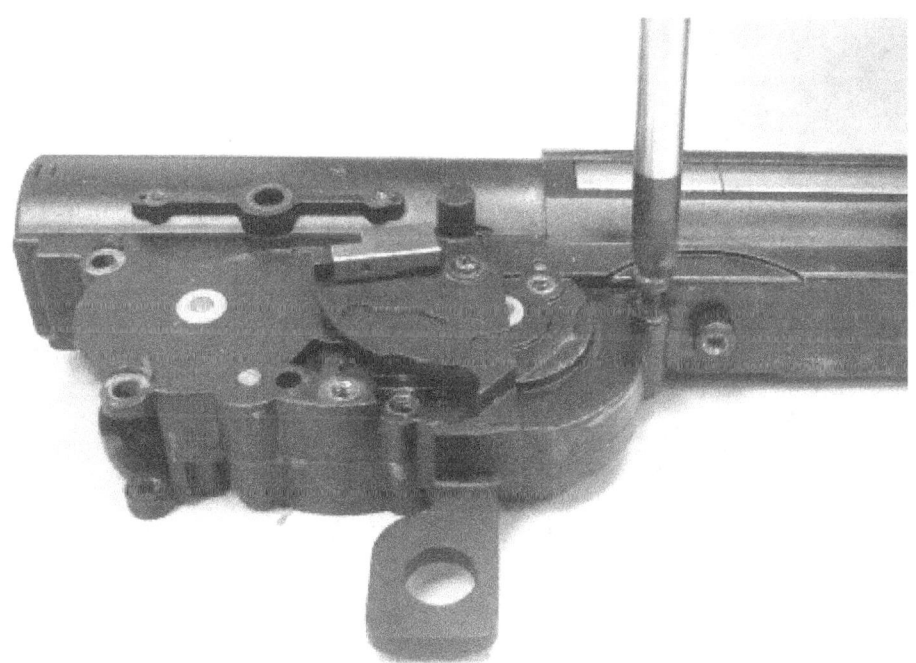

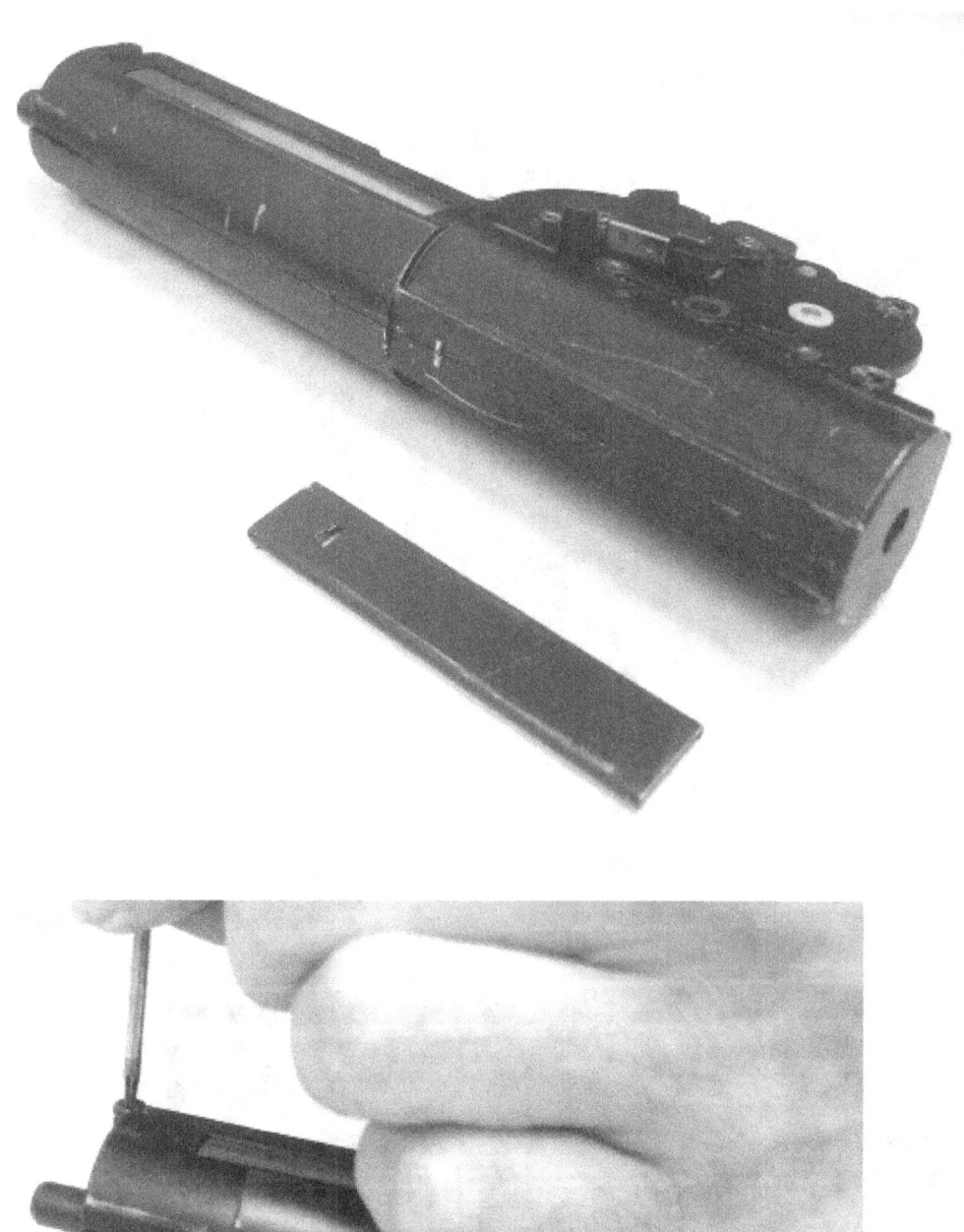

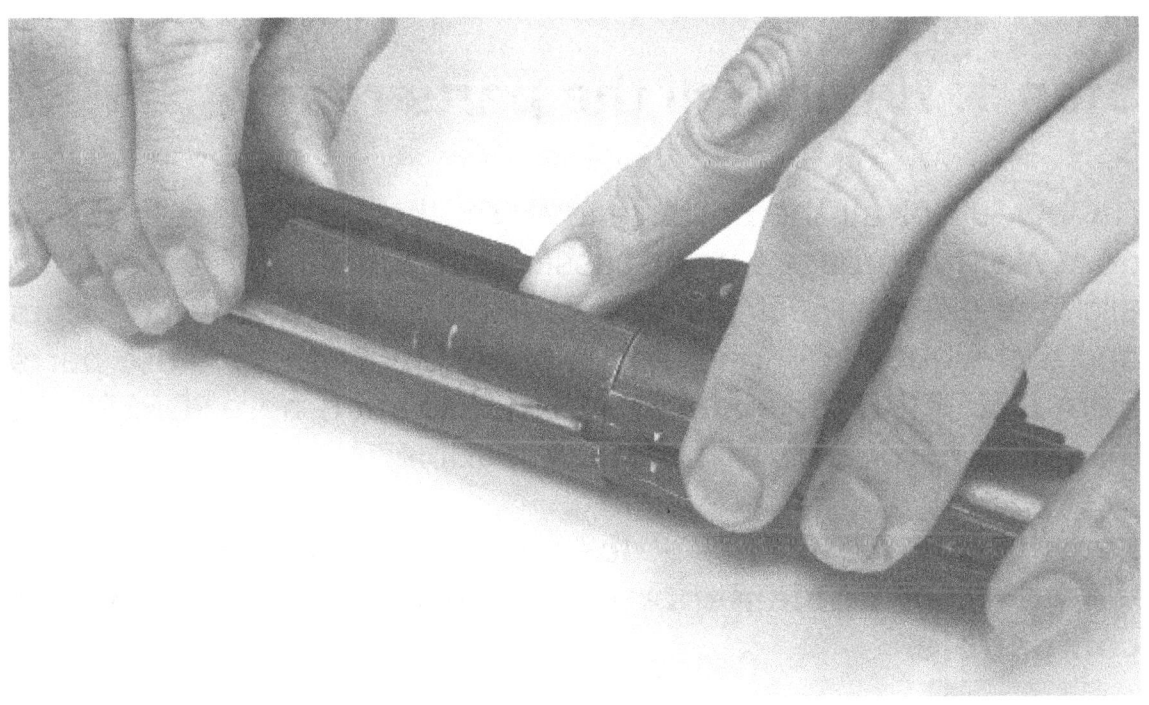

Step 1: Having all the parts ready

Have you got all the parts ready? You better ensure that you do... We try to base our discussion on standard TM mechbox architecture as much as possible even though the concepts and techniques are universally applicable. In fact we use parts from DEEPFIRE for extensive demonstration here. DEEPFIRE mechbox parts are 100% TM compatible.

Even though DEEPFIRE does not officially support full V7 gearbox implementation, many of their parts are compatible.

You want to know that M14 has a long inner barrel so it needs a cylinder of a full volume. The V7 cylinder is slightly longer than the "standard" cylinder.

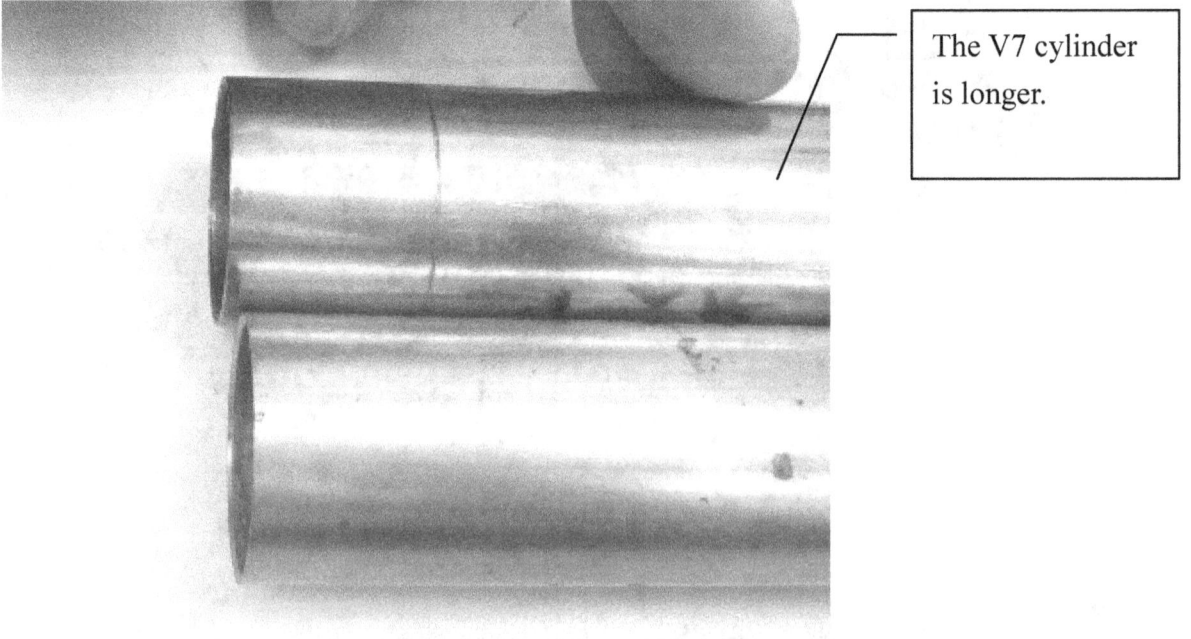

The V7 cylinder is longer.

Tappet plate for V7 is supposed to be model specific. On the other hand, almost all piston heads share the same size. In fact all V2 compatible pistons and piston heads can work fine in the V7 mechbox. Ball bearing piston head like the DEEPFIRE one can increase spring compression and reduce stress imposed by the spring on the gears. You want to attach the piston head to the piston early since they are supposed to function as one single piece:

Do not mix up the regular piston with the longer PSG-1 version. The PSG-1 piston is much longer.

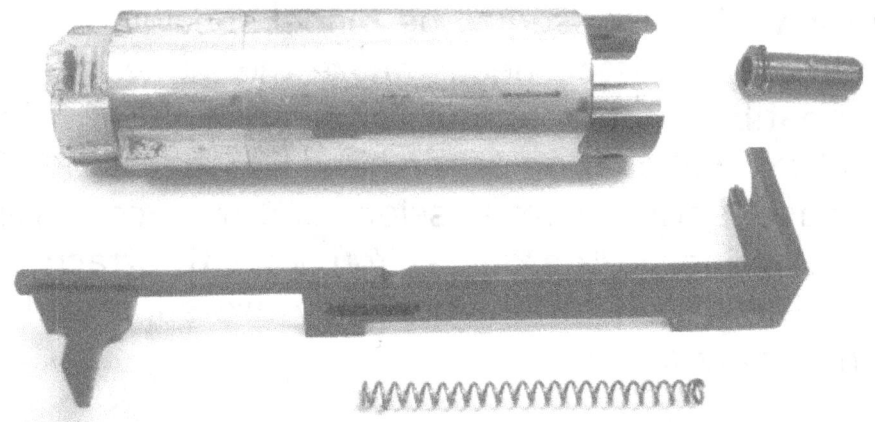

Note that the cylinder head, the tappet plate and the air nozzle are all V7 model specific.

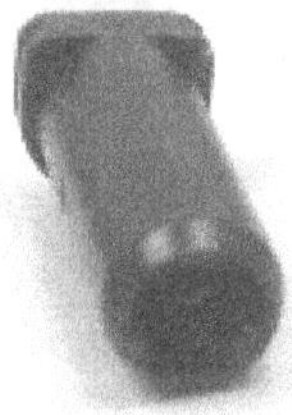

Specially shaped air nozzle. The nozzle opening is not exactly flat:

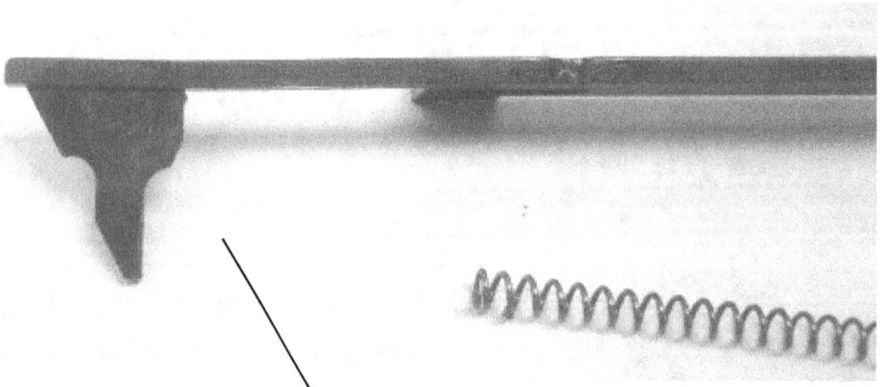

The tappet plate is similar to (but not an exact replica of) the V6 implementation.

The shell

For a high ROF setup, whether or not the shell has been reinforced does not make a difference in terms of the various assembly and troubleshooting processes. The TM V7 mechbox shell is good enough for up to M120 under proper use, while China-made shell can stay alive for up to M110 at the most.

For a high FPS setup, the M120 limit must be carefully accounted for. In fact, if you need high FPS performance on a TM mechbox there are TWO things that you must do:

1, anything over M120 / PDI 190% can crack the shell upon frequent full auto firing, so unless you change the shell you have a practical limit on the spring that you can use. Systema springs are generally more stressful for the gears so try to avoid them.

> Warning: I try not to recommend Systema M series spring because they are hard and long and are highly stressful for the gears. PDI springs (the 170% variant and the 190% variant in particular) highly recommended.

2, TM (and compatible) mechbox shell takes 6mm bushings. 6mm ball bearings may fit but a little bit of minor filing of the shell may be necessary.

>  NOTE: Surprisingly, the stock TM plastic bushings can handle a M110 spring in our test unit without breakage after about 9000 rounds in the field. Still, I wouldn't recommend that you use plastic bushings for good – just change them.
>
> Marui plastic bushings are more durable than the China-made plastic bushings. The Marui bushings are made of Nylon, while the china clones are made with cheaper resin material.

If you are to stay within a FPS of 320 or so it is recommended that you use ball bearings rather than metal bushings (ball bearings are more energy friendly). If you will go over this power limit, there are 2 recommended solutions:

i, use 6mm metal bushings, which would last very long and would cost less to setup.

ii, use 7mm (or even 8mm) ball bearings – you'll need to enlarge the bushing holders on the shell to accommodate 7mm ball bearings (the use of dremel tool would therefore be required –enlarging the holes from 6mm to 7mm is nothing technically sophisticated but you just have to practice a lot and be careful not to over dremel).

>  Warning: If instead of bushings you want to use ball bearings, 7mm bearings are okay for a PDI 170% but 8mm bearings would be recommended for PDI 190% (that means further dremelling would be necessary).

6mm ball bearings are generally not recommended for any FPS upgrade. Any thing over 300fps may be risky for 6mm ball bearings.

FYI, the DEEPFIRE guys disagree with this. They said if the mechbox configuration is done properly and if the ball bearings are made in Japan, M120 is perfectly safe for them.

China made clones often have incorrect and inconsistent sizing here… A 1mm difference can break the entire setup pretty fast!

The internals

As previously said, except for the piston, most components are V7 specific.

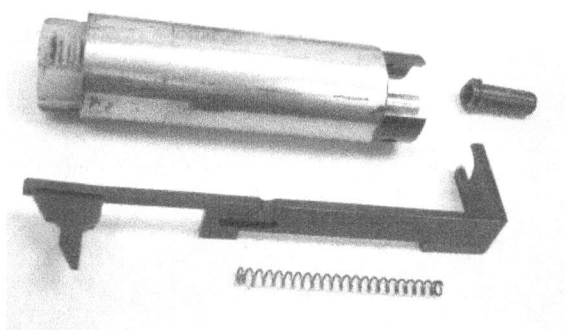

The gears are also different due to different size and gear ratio:

On the left: the V7 middle gear.
On the right: the standard V2/V3 middle gear.

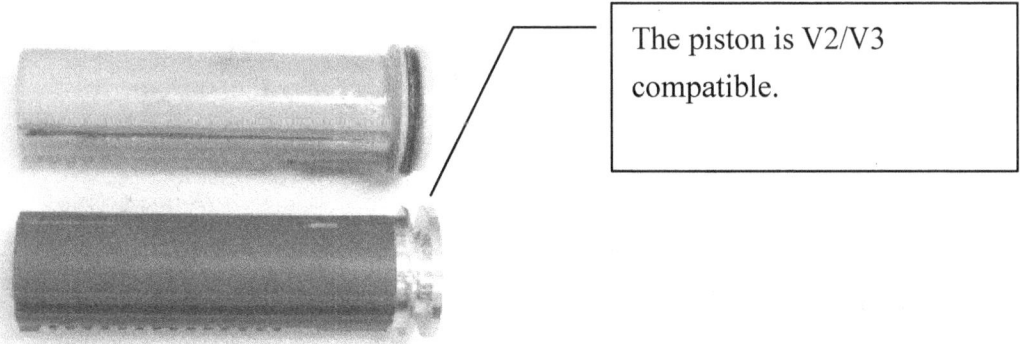

The piston is V2/V3 compatible.

You should check for air leakage of the cylinder by blocking the cylinder nozzle and trying to push in the piston. The easiest solution here is O ring replacement.

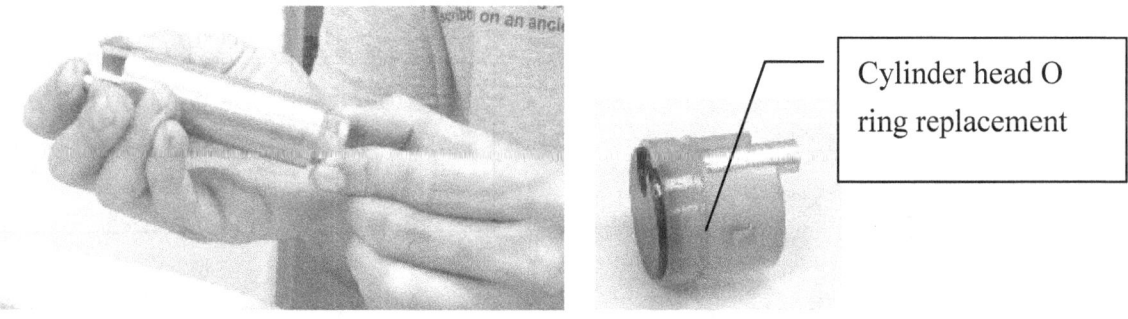

Cylinder head O ring replacement

Whenever you plan for replacing the nozzle, the cylinder or the piston, first check for air leakage. Strong resistance indicates relatively good air tightness.

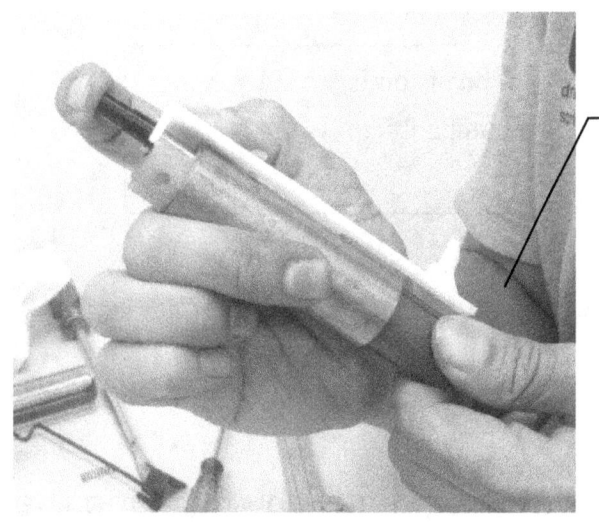

Proper air seal is the best way to improve performance cost effectively.

The electric switch assembly is not the same as the V2/V3 implementations even though the theory is identical.

Poor soldering work can lead to contact and other problems. Make sure the wires are properly soldered.

Upon frequent use these plates can deform. They are not expensive so replacement would be the best solution.

I strongly suggest that you first get all the necessary soldering works and wire connections done before putting the switch assembly into the mechbox. Depending on the gun model and the planned location of your battery, you'll need to plan the length of the wires for battery connection carefully. You must also plan the length of the motor wires – remember, the wires need to be properly routed inside the piston grip before reaching the motor (to avoid getting in the way of the motor shaft and pinion), therefore it is always preferable to keep the motor wires a little longer (only a little longer) then required.

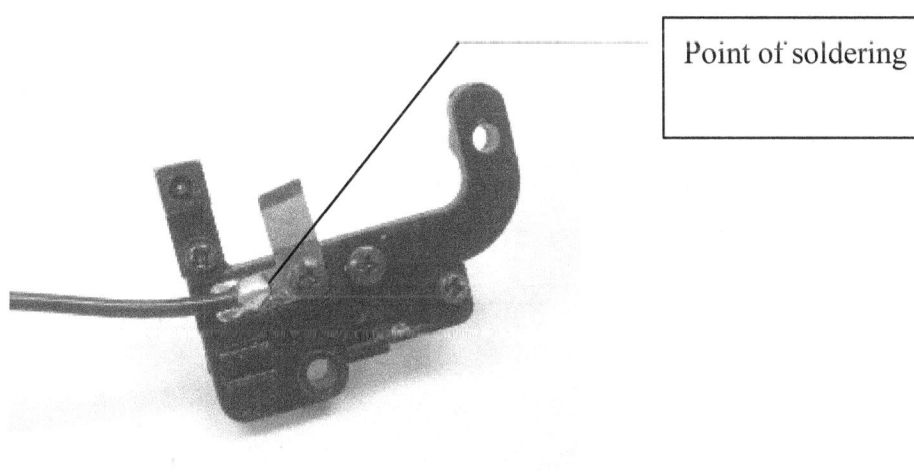

Point of soldering

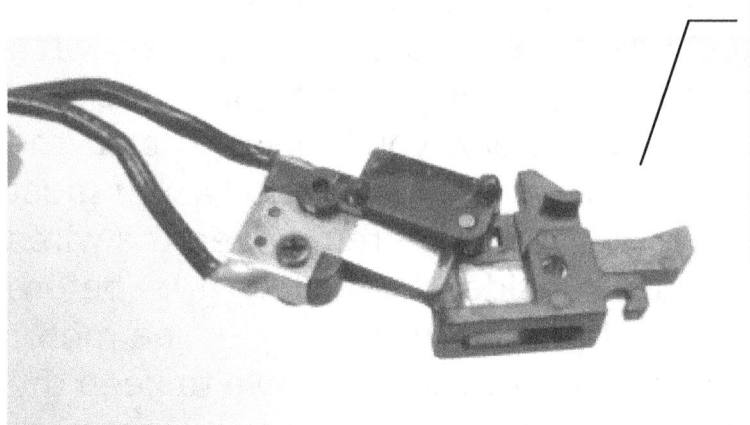

The operating theory of the electric switch unit is similar to that of a V2/V3 mechbox.

It works like this:

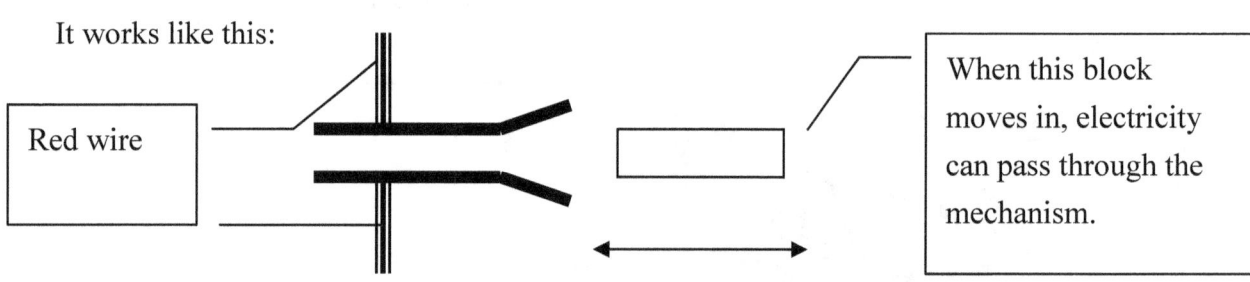

Red wire

When this block moves in, electricity can pass through the mechanism.

If you need to extend or patch the existing wires...

The goal of soldering is to join electrical parts together for forming an electrical connection. Basic supplies needed for proper soldering include a soldering iron (the prong of metal that heats to a specific temperature through electricity), the soldering wire (an alloy of aluminum and lead), and a cleaning resin called flux that ensures the joining pieces are incredibly clean (by removing all the oxides on the surface of the metal that would interfere with the molecular bonding, allowing the solder to flow into the joint smoothly).

The first step in soldering is cleaning the surfaces (including the iron tip itself). They must all be clean and free from contamination. Then, you may melt flux onto the parts to be joined. The parts should both be heated above the melting point of the solder but below their own melting point with the soldering iron. When touched to the joint, this precise heating can cause the solder to flow to the place of highest temperature and makes a chemical bond. Do keep in mind that too much solder is an unnecessary waste but too little may be insufficient for supporting the component properly.

Solder melts at around 190 degrees Centigrade. Such a temperature is hot enough to inflict a nasty burn. Be extremely careful when you do your soldering work. Also, do your soldering work in a room with good air circulation. Soldering does release toxic fumes.

Always try to keep the wires short as longer wires often induce much higher internal resistance.

Ultra FPS Technote

Your best bet is to use AWG 16 wires for everything. The stock TM wires may not be able to sustain the heat generated under the current draw produced by a high FPS setup. The China made wires would be even worse.

As will be discussed later we need to use strong battery power to pull an upgrade spring real hard. My advice is that you should forget about small cells entirely if possible when using PDI 170% or above. Use regular full size cells with at least 2000mah, 9.6V. You don't need 10.8V unless you go to PDI 230% or stronger.

Also pay attention to the motor you use.

AIP (Army International) produces VERY HIGH PERFORMANCE short motors that can be deployed on the M14. Its HS50000S has a RPM of 50000, while the HS40000S High Torque has 40000!

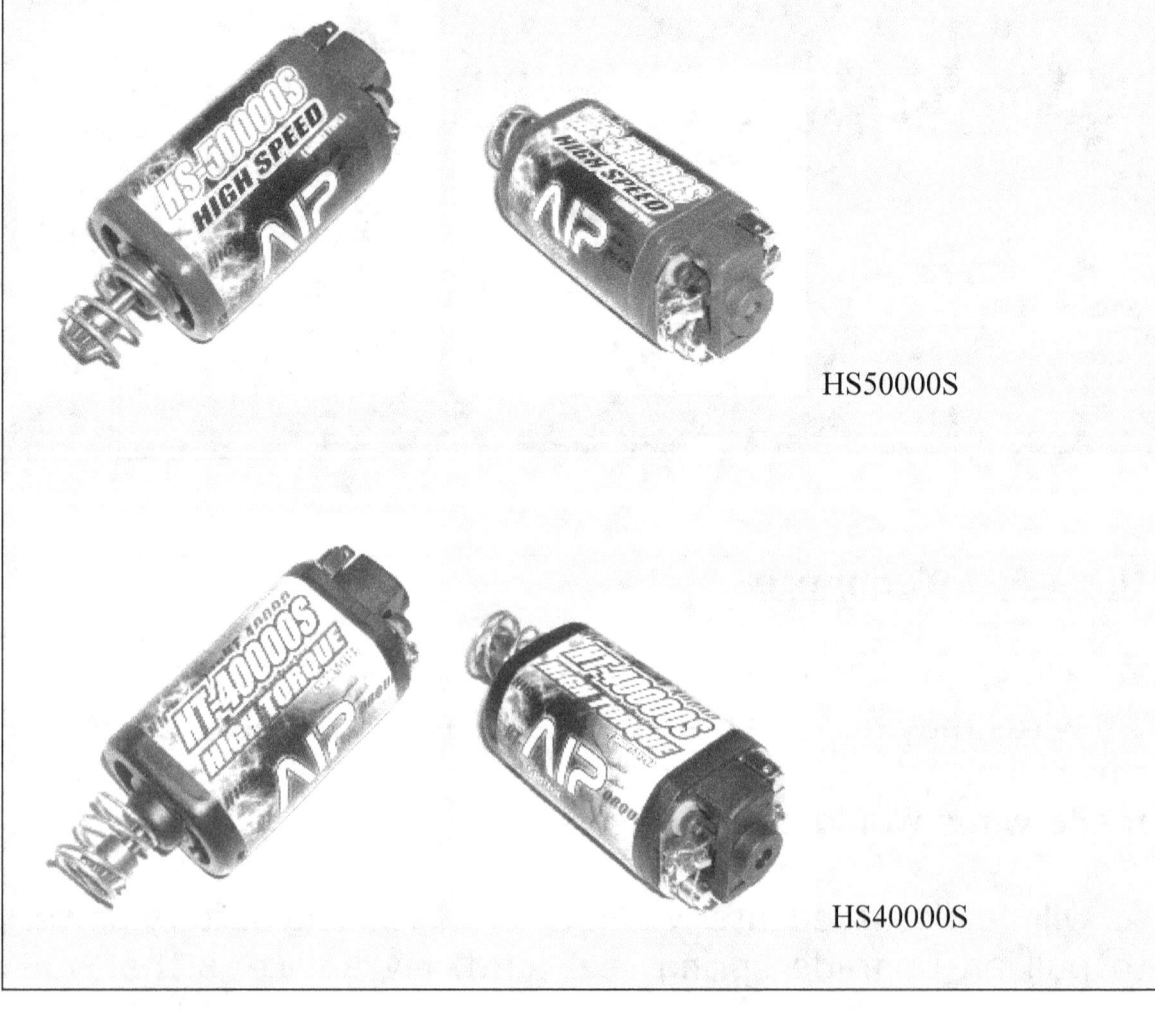

HS50000S

HS40000S

NOTE: EG1000 offers higher throughput than EG700 theoretically. In reality, there are many factors to consider for determining the actual throughput, and the difference on paper may produce benefits that are too marginal to be of any significance.

Different motors have different operating characteristics, that with different power source they may produce different RPM and different toque at different loads. And the efficiency does not have to be linear (refer to the example below):

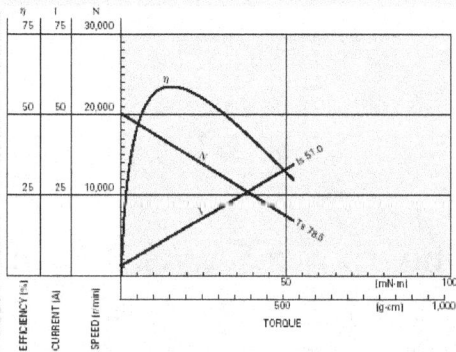

We do not want to go into details on motor technology. What we want to say is that, visual difference is minimal between most mass market motor models currently being offered in the market. If you want to see significant difference in torque or ROF, get a real high end motor such as the AIP HS50000S or change the gear ratio. The V7 gear ratio is already optimal so motor replacement would be the only viable choice.

As a new comer the AIP HS series is a surprise performance-wise.

Do note that proper motor break in can improve overall performance. Refer to the special technote at the end of this book for further information.

High ROF Technote

Again, your best bet would be to use AWG 16 wires for everything. The stock TM wires are just too thin to sustain the heat generated under the current draw produced by a high ROF setup.

For high ROF at a FPS below 320, you want at least a 8.4V pack with 3000mah or above. For higher FPS without sacrificing ROF, 9.6V 3000mah pack is preferable. If you have no space to accommodate full size batteries but only smaller size cells, use the Intellect 1400mah small cells to form your custom made 10.8V or 12V pack. The GP1100 small cells are marginally okay, but the Sanyo 600mah would be way too insufficient. Try to keep the wires short since longer wires often induce much higher internal resistance.

Also pay attention to the motor you use. EG1000 runs slightly faster than EG560. If you are using a Marui mechbox, then your choice is limited to EG560 though (unless you manually modify the shell).

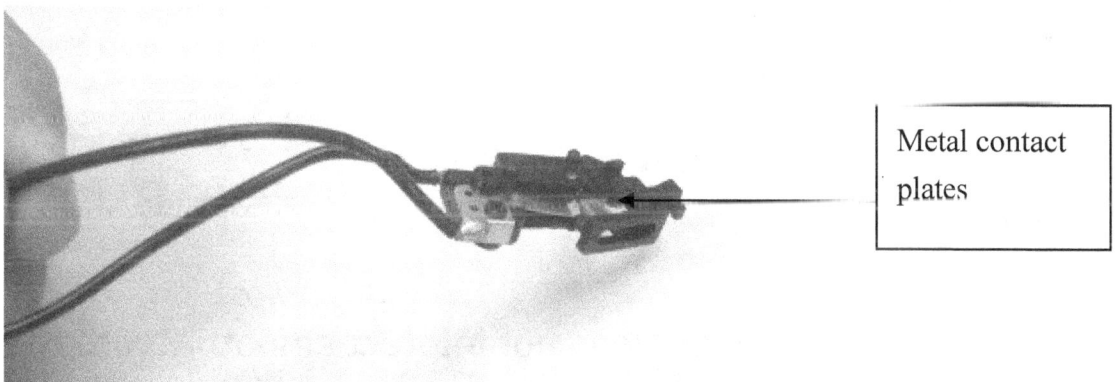

Metal contact plates

Be very gentle with the metal contact plates found on the switch unit. Don't bend the two metal plates on the switch (see below) or improper contact with the switch plate will result, which may disable semi or full auto firing entirely (depending on how they have been bent).

FOR CHINA MADE CLONE: Pay attention to the front of the

tappet plate. It has to be exactly 90 degree or problems will occur along full auto firing (be very careful with those white color tappets):

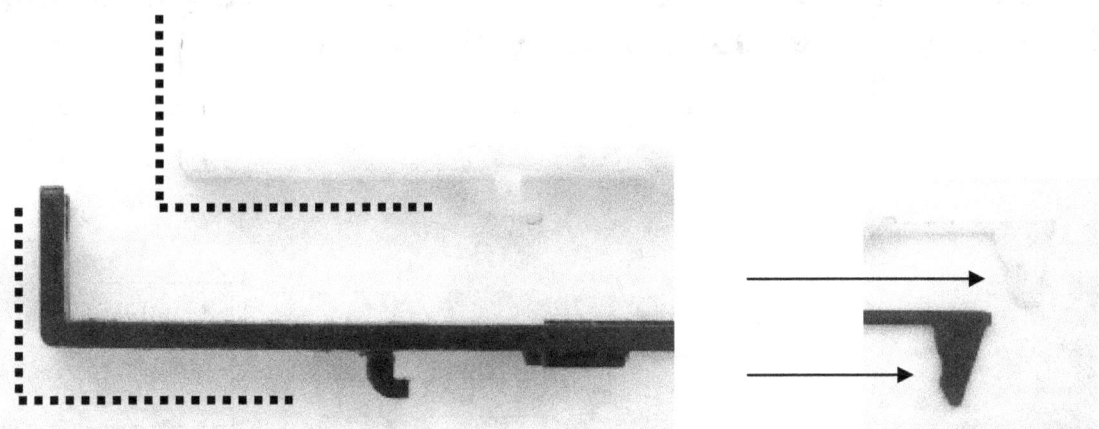

Note the tail. If the shape does not interact smoothly with the delayer (if you use one) then a little bit of filing would be necessary. Also make sure it can hold the air nozzle securely. This is not always the case if you mix and match parts of different brands:

> **Warning:**
>
> **!**
>
> Tappet plate breakage could be possible when you are running a high ROF setup together with the use of a sector gear delayer. You can't do much in this regard – without a delayer you MAY misfire all the time. Just make sure you buy a good quality delayer that will less likely produce troubles. A brand with the name Element is

pretty good at producing small mechbox parts such as delayer and small springs.

High ROF Technote

Below shows the spring that hooks the tappet plate to the shell. You may need to get it replaced more frequently under a high ROF setting as it will be under constant high stress. I would suggest that you replace this spring every 40000 shots.

Both the TM version and most China-made clones use plastic bushings. You want to take them out and replace them with 6mm metal bushings. You want to slowly hammer the metal bushings onto the shell:

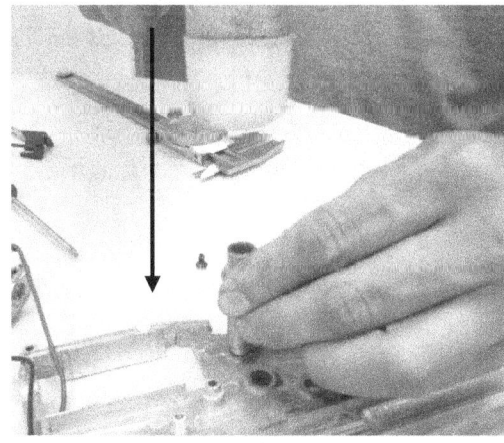

AirsoftPRESS (Hong Kong).

Shimming adjustment should be done on the gears one by one. First the middle spur gear, then the bevel gear, and finally the sector gear.

Ultra FPS Technote

A recommended approach to ultra FPS upgrade is to do it step-by-step. First try out a relatively mild upgrade - one that won't hurt much even if you make minor mistakes. A PDI 170% spring with metal bushings and all stock gears/piston will just be fine. Add 2 or 3 ring spacers to the stock spring guide and you should be able to safely go close to 380fps. Run this configuration for several games and if things go okay you may try the next step up.

If you experience frequent gun lockup (the motor fails to complete spring compression) at this power level something is wrong with shimming (or weak motor power due to poor wiring or weak battery...etc). Almost all EG560 compatible motors in the market can handle PDI 170% flawlessly under proper configuration, unless something gets stuck inside (sometimes removing the sector gear delayer may help). The problem with the cheaper clones is that they won't deliver enough torque when battery power drops below a certain level... the newer motors are supposed to be better.

If your wires are running hot, replace them with AWG16 wires ASAP and then switch to Dean plugs.

Ultra ROF Technote

There are crazy people out there who build high ROF mechbox though extreme teeth removal. They like to:

- cut several plastic teeth away from both sides of the piston (usually 3 from the side of the piston head, then move the metal tooth forward; and one from the rear, right after the biggest tooth)
- cut some teeth away from the sector gear (those teeth that were in touch with the piston – usually cut 3 away)

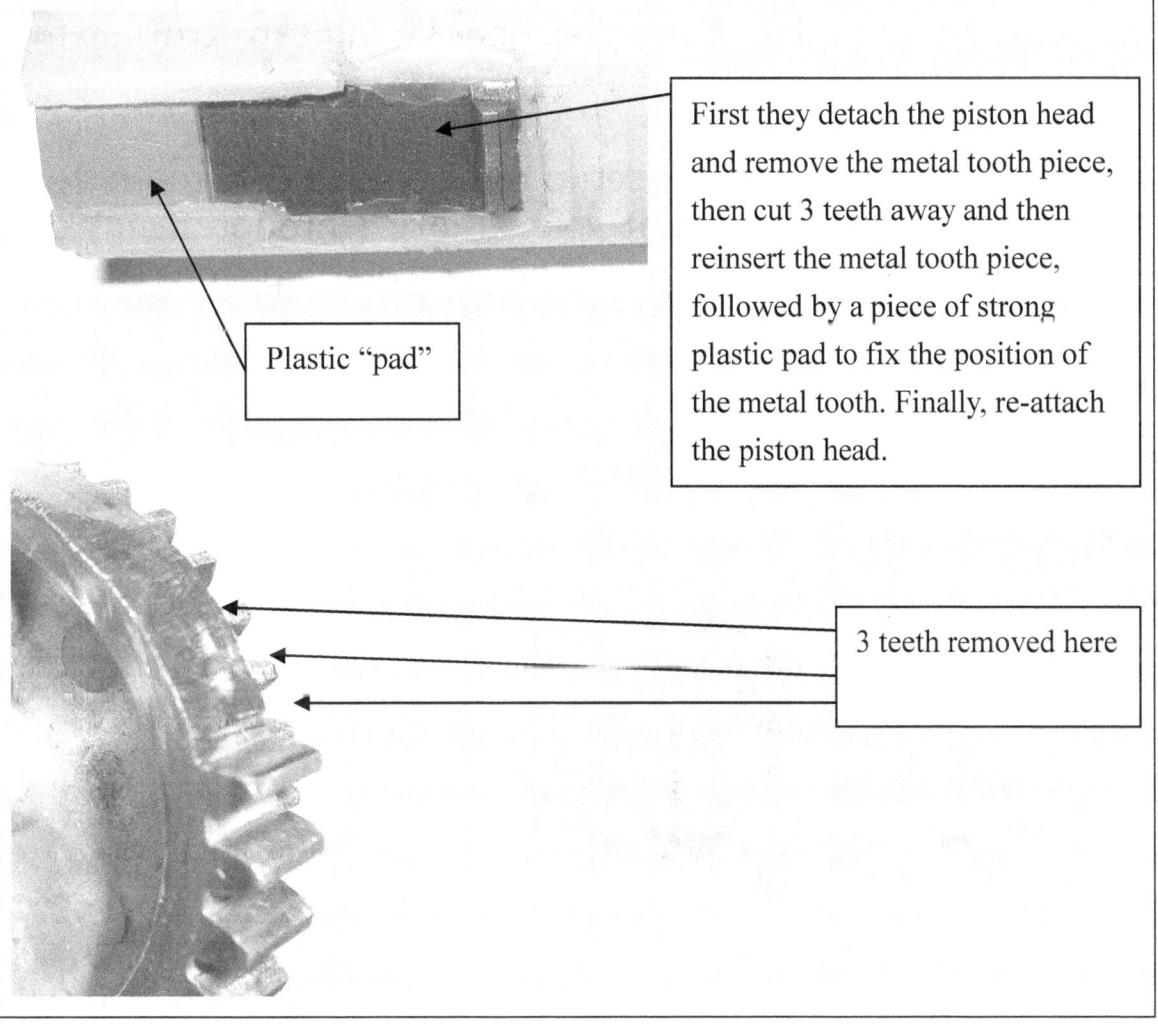

Plastic "pad"

First they detach the piston head and remove the metal tooth piece, then cut 3 teeth away and then reinsert the metal tooth piece, followed by a piece of strong plastic pad to fix the position of the metal tooth. Finally, re-attach the piston head.

3 teeth removed here

The primary purpose of teeth removal is to prevent stripping of the piston teeth. When ROF is too high the sector gear may attempt to pull the piston before the piston is back to the ready position.

Cutting teeth away would NOT increase ROF. It would, however, help in preventing gear breakage.

We do not recommend this radical method to beginners as it would most likely lead to insufficient compression as well as corrupted gear timing. Corrupted gear timing can screw up the whole mechbox...

A less aggressive but safer setup is to remove ONE tooth from the piston. We suggest that you remove (through filing) the second tooth counting from the rear side of the piston (the one right next to the "big tooth"):

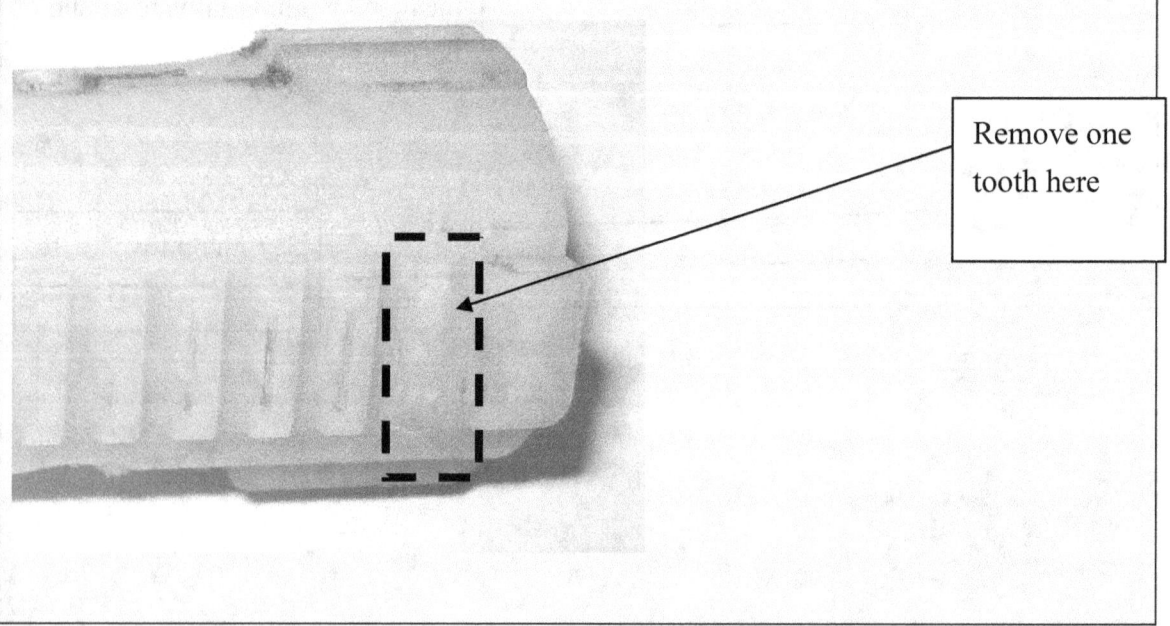

Remove one tooth here

The rationale behind this decision is simple. When doing ultra high ROF you get your piston or your sector gear stripped because the sector gear starts its next turn before the piston has fully "rebound". In a normal setup the piston is pushed fully forward for hitting the cylinder head before getting pulled by the sector gear. If ROF is too high, the piston gets pulled again before it hits the cylinder head (this can happen with the use of a stock spring – the stock TM spring is too soft, not strong enough to push back the piston in time). That means the first tooth of the sector gear may crash into the first several teeth of the piston (counting from the rear side of the piston). By removing one tooth from that risky area such crashing may become less likely.

Plastic pistons are easier to modify and are the ideal candidates for testing different high ROF approaches. Do note that the above way of teeth removal is simply one variation of the many removal schemes. Different teeth removal scheme is required for different setup requirements. They also work differently on gears of different makes. We will deal with this in depth in our Advanced AEG Upgrade title. The picture shows an example teeth removal scheme suggested by the guys from DEEPFIRE.

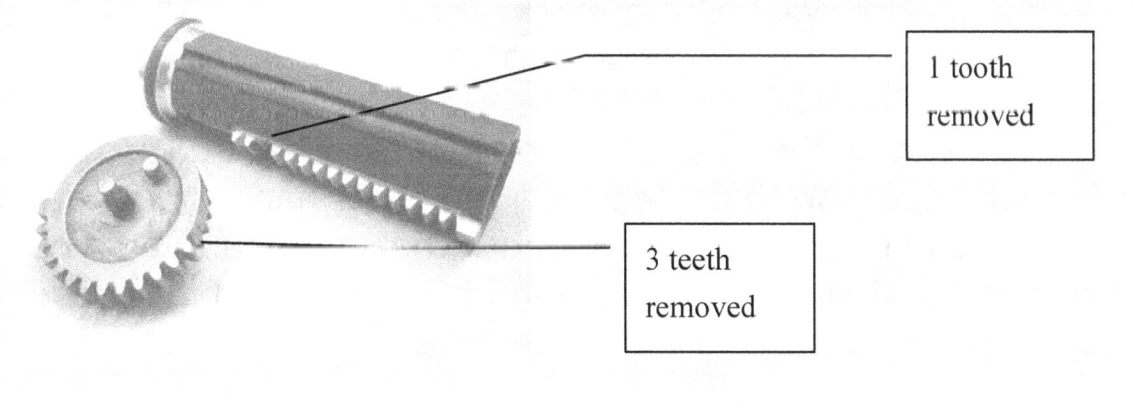

1 tooth removed

3 teeth removed

When dealing with third party replacement of the cylinder, the piston, the air nozzle, the spring guide and the spring, you need to pay particular attention to the issues below:

1, Does the air nozzle have an inner dimension that allows a smooth fit with the cylinder head nozzle? A loose fitting will fail to seal air, while a tight fitting may lead to air nozzle breakage or jamming.

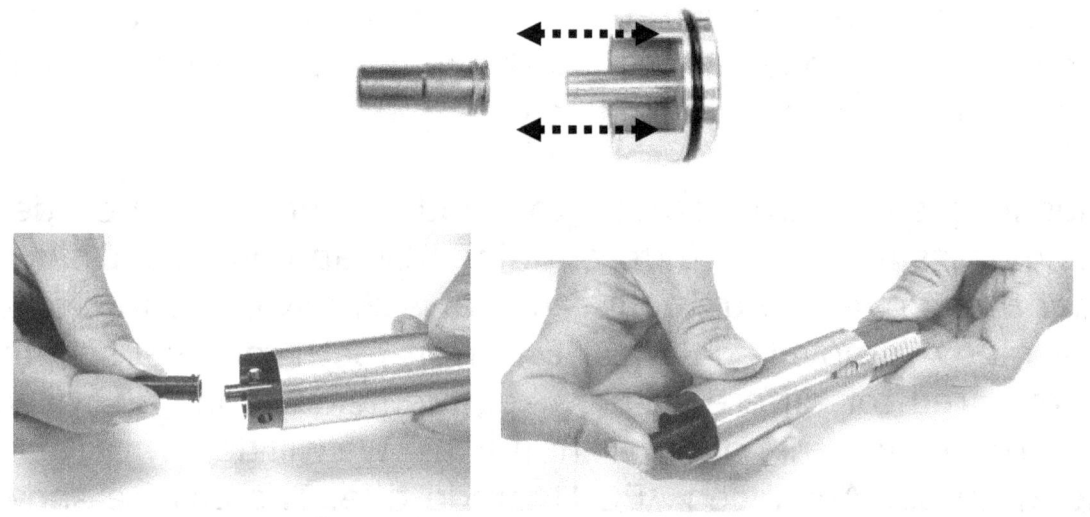

You want to test air tightness early. With one finger blocking the air nozzle you want to try pushing the piston and feel the resistance. Fix any leakage before installing the mechanism onto the shell.

Another thing, is the nozzle of a correct length? A nozzle that is too long can lead to misfeed. On the other hand, one that is too short can lead to FPS drop. Check the specification of your gun model to find out the exact length needed, and make sure your nozzle doesn't go 1mm more or less.

> **Warning:** If your gun is a made-in-China clone the air nozzle issue must be properly addressed before you move on. The made-in-China air nozzles are ALWAYS not accurately shaped and should be avoided whenever possible.

Ultra FPS Technote

The sector gear delay loading chip (the delayer) is to be installed on the sector gear so when it pulls the tappet plate it can pull it backward a tiny bit further and make it stay there a little longer, giving enough time for the BBs to feed. It is helpful when you are having a high ROF setup or that you are using heavier BBs (such as 0.25g BBs). HOWEVER, it may increase the workload of the motor by a bit. Frankly, you don't really have to install a sector gear delayer if your desired ROF is not going to change much.

Ultra ROF Technote

The delayer is helpful when you are having a high ROF setup or that you are using heavier BBs (such as 0.25g BBs). **This is a better solution than cutting the air nozzle short*!** And don't forget to use high quality mags to avoid misfeed. Stock TM mags are perfect. Those cheaper china made mags are no good and are often the major source of dry fire...

Do remember, the delayer must be tightly installed and fixed into position. If it goes loose it can crack the sector

gear and the tappet plate. Below shows the correct way of installing the delayer (if you insert it the other way around you can break the tappet plate):

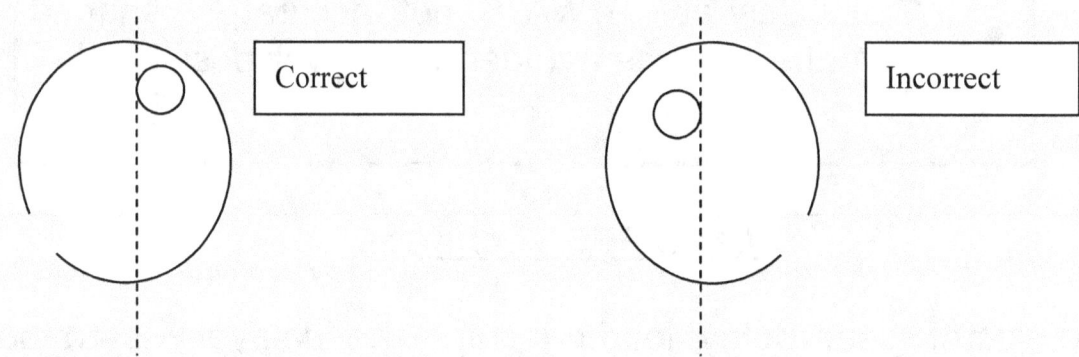

* For your reference:

With a stock air nozzle, when you use BBs that are heavier than 0.2g each, frequent misfeed may result. You may work around this by shortening the air nozzle a little bit (0.5mm max just to be safe). A better yet solution is to install a sector gear delay loading chip so that there is enough time for the heavier BBs to get properly loaded into the chamber.

If you have your air nozzle shortened by 1mm+, you may start to see FPS decrease plus occasional jamming when the hi-cap mag is almost running out of bullets. If you get your air nozzle shortened by 2mm+, you may see multiple bullets coming out of each single shot! This is because the nozzle is now too short to stop the next few bullets from squeezing into the chamber. This nozzle is hardly useable because your gun will jam like hell when doing full auto.

> Warning: Not all delayers can fit into your sector gear. Always test fit them beforehand – a loose delayer can screw up the tappet plate instantly. We have tried out the Element delayer – it has an okay fit with the TM sector gear.

2, If your cylinder head is not of the one-piece type, does the cylinder nozzle wobble? If it does, fix it (instant cement will do) or get a replacement before moving on.

> Warning: A cylinder nozzle that wobbles could indicate some serious quality problems that you should be aware of... TM cylinders are usually free of this kind of problem but the China-made cylinders are not.
>
> I always prefer a one-piece design here. Like the DEEPFIRE one, it is way more solid. Installation wise it is easy and trouble free. For V7, however, choice is limited......
>
> The traditional multi-pieces design looks like this:

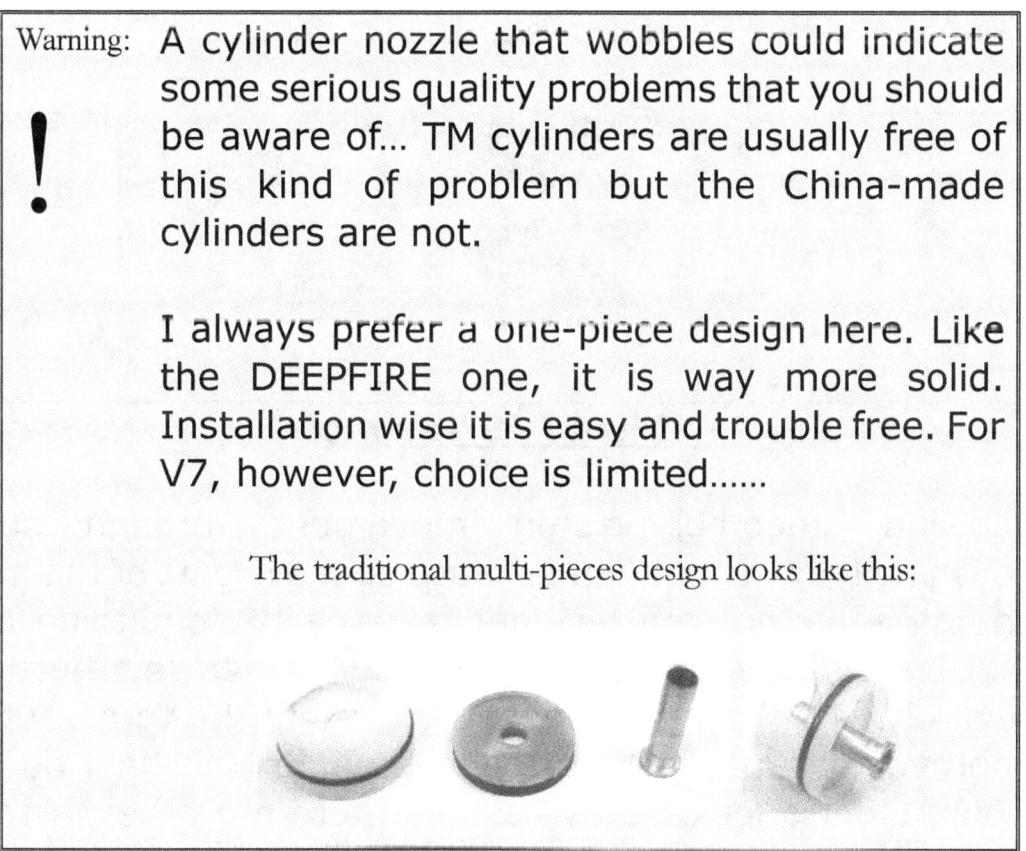

3, Test fit the piston with the cylinder. When you move the piston inside the cylinder you should feel a certain level of resistance. If there is no resistance at all, the piston or the piston O'ring is no good (it is allowing too much air leakage, which can result in real serious performance drop). On the other hand, if there is too much resistance, the spring will have a hard time pushing the piston and will lead to both FPS drop and ROF drop. Based on our experience, the stock TM piston is one of the best in terms of smooth movement here.

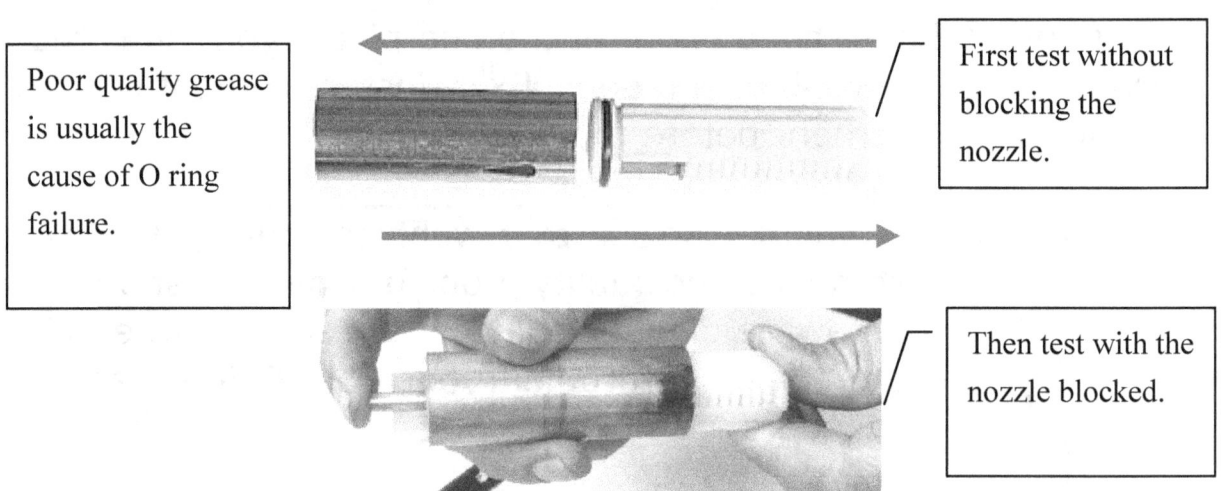

Poor quality grease is usually the cause of O ring failure.

First test without blocking the nozzle.

Then test with the nozzle blocked.

Ultra FPS Technote

Allowing smoother piston movement through surface polishing MAY benefit FPS because speed is part of the impact energy equation. You hit your target with something heavy that travels at a high speed and your target gets hurt more severely. The thing is that the benefit in terms of FPS improvement may not be too noticeable (the final outcome would differ a lot case by case as there are other factors at work). **ROF-wise YES you will see a difference!**

You may reduce friction here by polishing the inner side of the cylinder. You don't really need to sand it. Just wax the inside several times. This can allow smoother piston movement and increase the life span of the piston head O ring.

Also pay attention to the side stripes of the piston. We found that certain made-in- China pistons have side stripes that are not deep enough to fit with the tracks on the mechbox shells (you will have a hard time closing up the two halves of the shell back together). Again, you can perform test fitting prior to formal assembly. You must be able to get the two halves of the mechbox shell perfectly closed and at the same time allow the piston to slide smoothly.

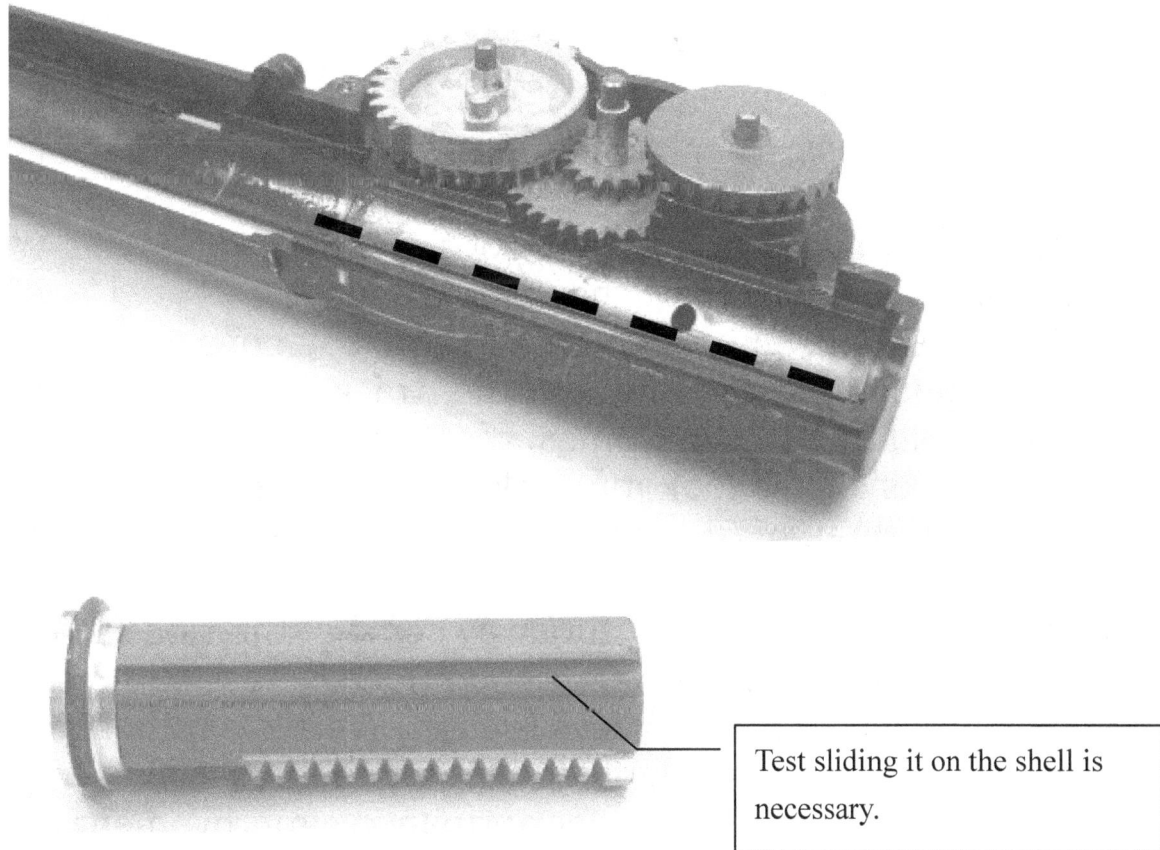

Test sliding it on the shell is necessary.

Ultra FPS Technote

You can reduce friction here by sanding the side tracks on the mechbox shell. These are what you will need:

- P400 Sand paper
- P600 Sand Paper
- P800 Sand Paper
- P1000 Sand Paper
- A bucket of water
- 2 rags
- Rubbing Compound
- Wax

Steps:
1. First use the P400 Sand Paper to sand in one direction a couple of times. Then wipe the tracks with a wet rag. Dry them.
2. Use P600 to do the same as above.
3. Use p800 with water do the same as above.
4. Use 1000 with water and do the same as above.
5. Use Rubbing compound on the tracks. Pass Rubbing compound three or four times, removing it every time. The side tracks should look much clearer as a result.
6. Pass wax three or four times and you are done.

On the other hand, you may cut away some of the side stripes of the piston to reduce overall friction (ideally, less surface in touch with the side tracks = less friction). This is a rather safe measure as long as you don't cut too aggressively. True effectiveness in terms of performance improvement is un-verifiable though...

You may also improve FPS by making the piston lighter (so it may travel faster). See the picture below and you'll know what I am talking about... It looks a bit crazy, but it works even though the FPS gain is quite marginal...

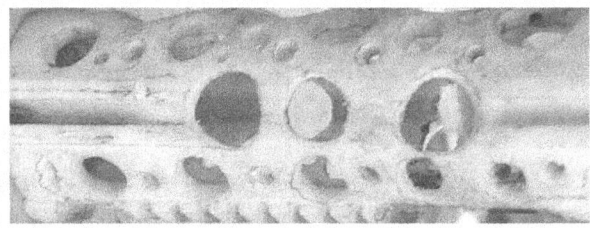

To get started safely without damaging the piston when drilling, start with a bunch of small holes that spread evenly across the piston body.

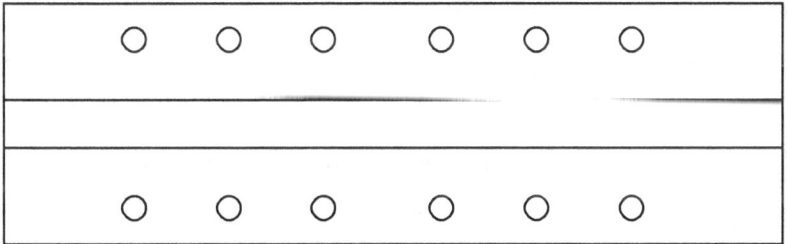

A question to ask is, will a lighter piston deliver less impact? In theory, yes, because the moving mass is less heavy (speed and weight both contribute to Impact force). However, a lighter piston can move forward faster when the compressed spring is released, so you get positive compensation here. Based on our test results a lighter piston is POSITIVE for both FPS and ROF, although FPS contribution is relatively less significant than ROF contribution.

4, Test fit the cylinder head with the cylinder. A loose cylinder head can allow serious air leakage. Pay particular attention to the cylinder O Ring – if the cylinder head is loose, try to see if you can replace this O ring with a larger one.

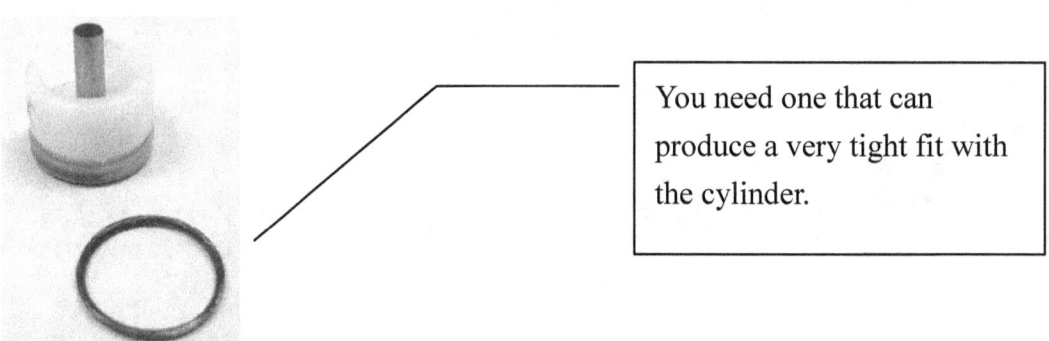

You need one that can produce a very tight fit with the cylinder.

Warning:

!

Don't confuse a piston head O ring with a cylinder head O ring. They are 2 different things. When buying a new piston head O ring, by all means take your current cylinder / piston set with you to the hardware store and try things out on the spot. You want an O ring large enough for a slightly tighter fit with the cylinder (so you can improve compression), but at the same time you don't want an O ring that is too large. An over-sized O ring will make it very hard for the piston to move inside the cylinder and will increase the load of the motor and the battery quite significantly.

Remember to also check the inner diameter of your replacement O ring. If it is too large (an inner diameter almost as large as the overall diameter of your piston head is way too large), it may fail to sit inside the "channel" of

> *the piston head during rapid piston movement. If part of it comes off, your mechbox will immediately jam.*

Applying grease around the piston head O ring is a measure primarily for ensuring smoother movement – with proper lubing the O ring can have less wear and tear (and can last longer). Most of the time you should not need to replace the O ring unless FPS drop becomes apparent. Do keep in mind, an O ring that fits with the cylinder tightly WILL wear faster due to more friction imposed on it during piston movement. **Poor quality grease can damage the rubber O ring over time so make sure you get some real good ones.**

> Warning: The cylinder head O ring is made of very thin rubber and is quite difficult to acquire from the local hardware stores (due to its unique sizing). If you don't have one handy, the best thing to do is to apply grease around it for better air sealing. Since the cylinder head does not move, in theory its O ring should not wear out by itself. Still, you may want to regularly check and reapply grease around it if necessary during your regular maintenance effort.

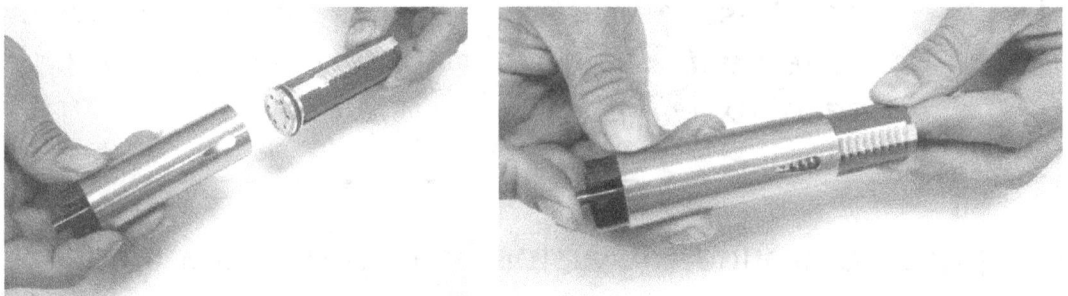

Test air tightness again. This time your focus is on the piston. Piston O ring replacement may become necessary in case of leakage.

> NOTE: FYI: When the barrel is too short, too much air from the cylinder will actually affect bullet stability. This is why you need an auxiliary port equipped cylinder so that "extra" air can be leaked out without affecting the bullet. However, when the barrel is a long one, you don't want any intentional air leakage through the auxiliary port or you risk pushing the bullet without sufficient air.

5, If you are using a custom made spring (don't get me wrong, I have nothing against custom made spring), before installation you need to test fit it both sides with the spring guide. Installation-wise the side with the tighter coils always go to the spring guide, but when performing a test fit you must do it both sides because along spring compression most part of the spring will eventually be in touch with the spring guide and if any part of the spring does not fit with the spring guide your mechbox will get locked up eventually.

Remember, a good fit should allow all parts of the spring to freely rotate along the spring guide (even when the spring is fully compressed). If the spring cannot rotate freely, something is wrong and jamming may occur.

Some springs are simply too large for some pistons. For example, many China-made springs would not fit into the DEEPFIRE piston.

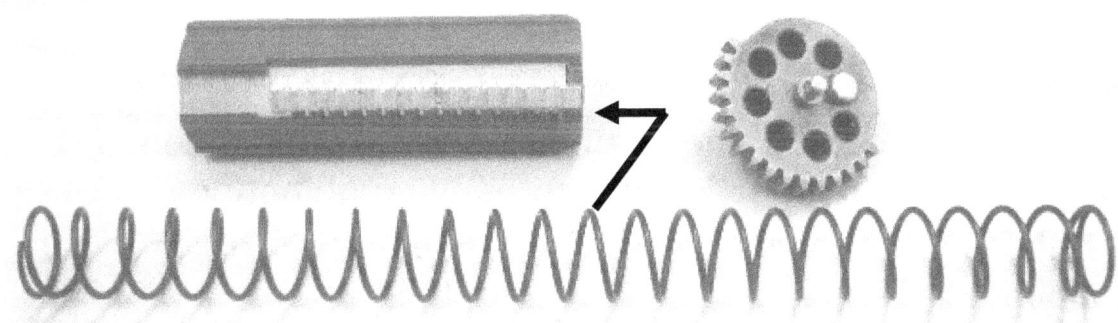

Ultra FPS Technote

A PDI 190% spring in your AEG may be quite close to breaking the law of your country (for example, in the UK where you cannot go over 1.3J something for AEG a PDI 190% is already way too much) - you better check before proceeding.

Keep in mind, if you are using a very long inner barrel you should not use a cylinder that has ports on it. A hole on the cylinder for intentional air leakage is good ONLY for short barrel setup as it allows the piston to get pushed back faster without disrupting air volume balance.

Tight bore barrel can increase FPS, but given the fact that bullets quality can vary greatly these days I would not risk using a tight bore (jamming can be a real headache on the field).

NOTE: You don't need a ball bearing spring guide and a ball bearing piston head combination at PDI 190%. However, installing these can reduce tension built-up and can improve compression a little further (plug the benefit of minor energy saving), which is ideal given the spring power limitation imposed by the stock shell.

High ROF Technote

Talking about spring, keep in mind it is incorrect to assume that a soft spring is better for high ROF configuration. A spring that is soft often fails to rebound to the full position, which may lead to gear stripping and other problems. For high ROF, you need a hard spring that can rebound fast enough. This is why you also need to use a strong battery – you pull the spring hard, and let the spring rebounds hard. Check out the spring shown below. I cut it out from a full length M120 spring. It is now much shorter than a regular spring but because it is hard it can sustain high ROF configuration without losing too much power (ROF 27 at FPS 320 under a 8.4V 3300mah pack):

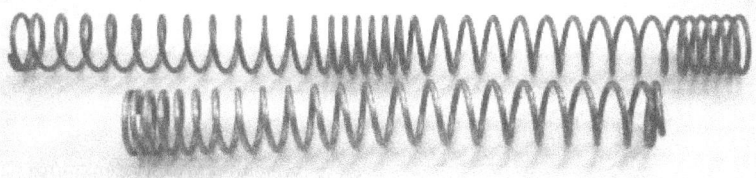

Below shows some springs of different lengths simply for your comparison purpose:

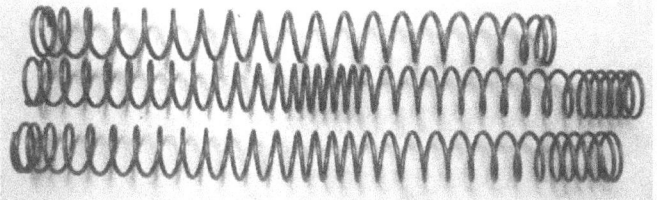

Springs with "thicker" wires are usually harder. Springs with thicker wires and more turns are of course way harder. Another thing – some springs may be too "large" diameter-wise for fitting into your piston – you just have to perform testing fitting in advance.

When choosing a spring guide or adding spacers, always assume full compression is taking place. If teeth removal is performed, full compression may not be possible. Still, to avoid possible breakage it wouldn't hurt to assume full compression upfront.

You have to keep in mind, significant teeth removal means there is less compression (less teeth to drive the piston backward) plus less air inflow (the piston travels

less means there is less space for accommodating air inflow). This is a tradeoff that is unavoidable. AND BESIDES, M14 IS SIMPLY NOT AN IDEAL CANDIDATE FOR A HIGH ROF SETUP.

The anti-reversal:

Note that some third party anti-reversal latches have an axle too small to fit onto the placeholder of the mechbox shell, thus resulting in wobbling. Test fit it early to determine if replacement is necessary.

This is where the anti reversal should be inserted.

High ROF Technote

Be expected to replace the entire anti reversal unit more often under high ROF (the tip of the anti reversal can get worn out very fast). At least replace the spring that supports it on every 35000~40000 shots.

You should keep enough spare washers in your tool box. You will often need to use them when changing gears (especially when switching to gears from a different manufacturer).

Gears

I must say that the stock TM sector gear is way too weak. Too easy to wear out under increased load. Replacement is highly recommended, but you may better off replace the entire set of gears rather than just a single gear. This is due to the concern on gear meshing.

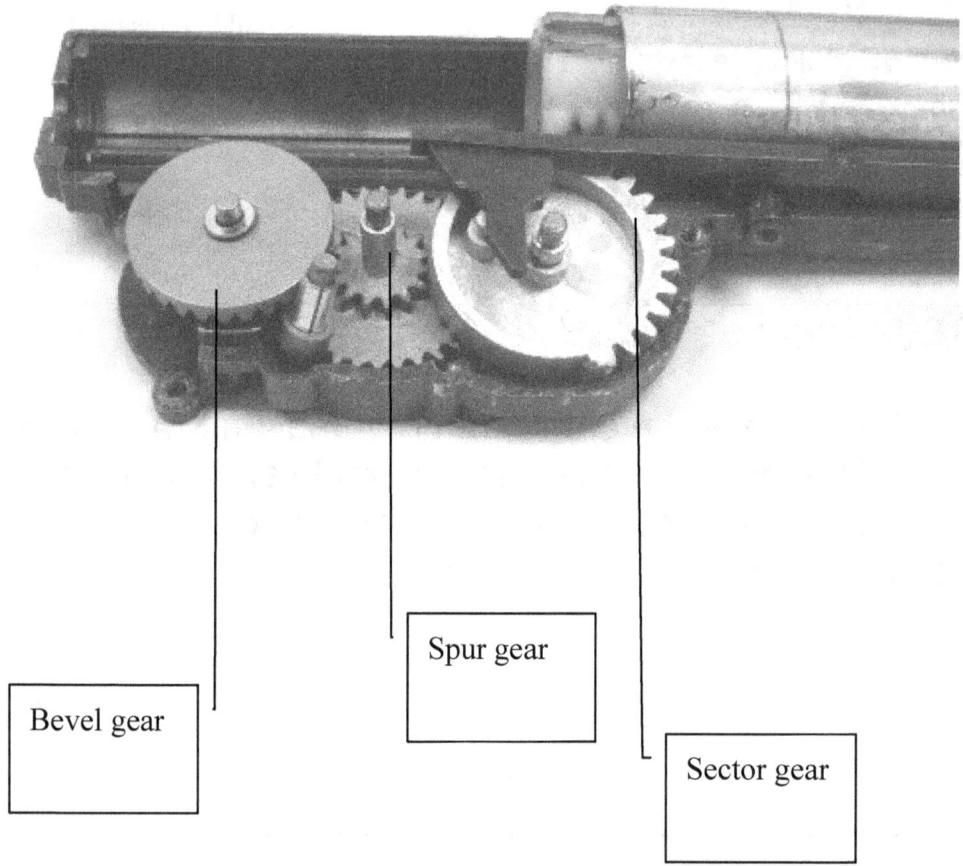

Bevel gear

Spur gear

Sector gear

Try to mesh the gears by hand and see if they fit together well. In fact it is always suggested that you buy all three gears from the same manufacturer to ensure the best possible fitting.

> **Warning:** Also check the diameters of the gear axles. Some gears may have axles too big to fit into the existing bushings. A good fit here is critical to the operation of the mechbox. Absolutely no tolerance on error here.

Ultra FPS Technote

You don't need to use high torque gears - these gears are for things well over PDI190%. In terms of motor, even the Chinese made EG700 replica would be good enough at this power level provided that proper wires and proper battery power (9.6V, high mah value) are deployed and maintained.

We repeatedly emphasize the importance of the use of proper wire. Remember, the motor has a metal shell to help dissipating heat, but the wires do not. The wires when overheated will refuse to work.

Mechbox gears can strip easily if:

- you go full auto all the way until an entire hi-cap mag is emptied
- your mechbox is running way too fast
- misaligned gears and improper shimming
- broken bushings

Gear stripping is a fact of live. Every mechbox will come across this problem – it is just a matter of time. Proper usage together with proactive maintenance measures can delay stripping. Open up your mechbox regularly (like every 10000 shots) and do a check up. Remove the dirt and other unknown particles that are accumulated inside the mechbox. Apply grease if necessary.

Metal bushings (OR ball bearings):

Not all bushings are of exactly identical dimension. Therefore, you have 2 issues to deal with here:

1, Compatibility with the gears: test fit the gears with the bushings and make sure that the bushings can comfortably accommodate the gear axles without much "play".

2, External compatibility: test fit the bushings with the mechbox shell. The bushings should NOT be allowed to spin on the shell. A tight fit here is important.

Standard bushing:

If the bushings are OK, then have them firmly inserted into the mechbox shells.

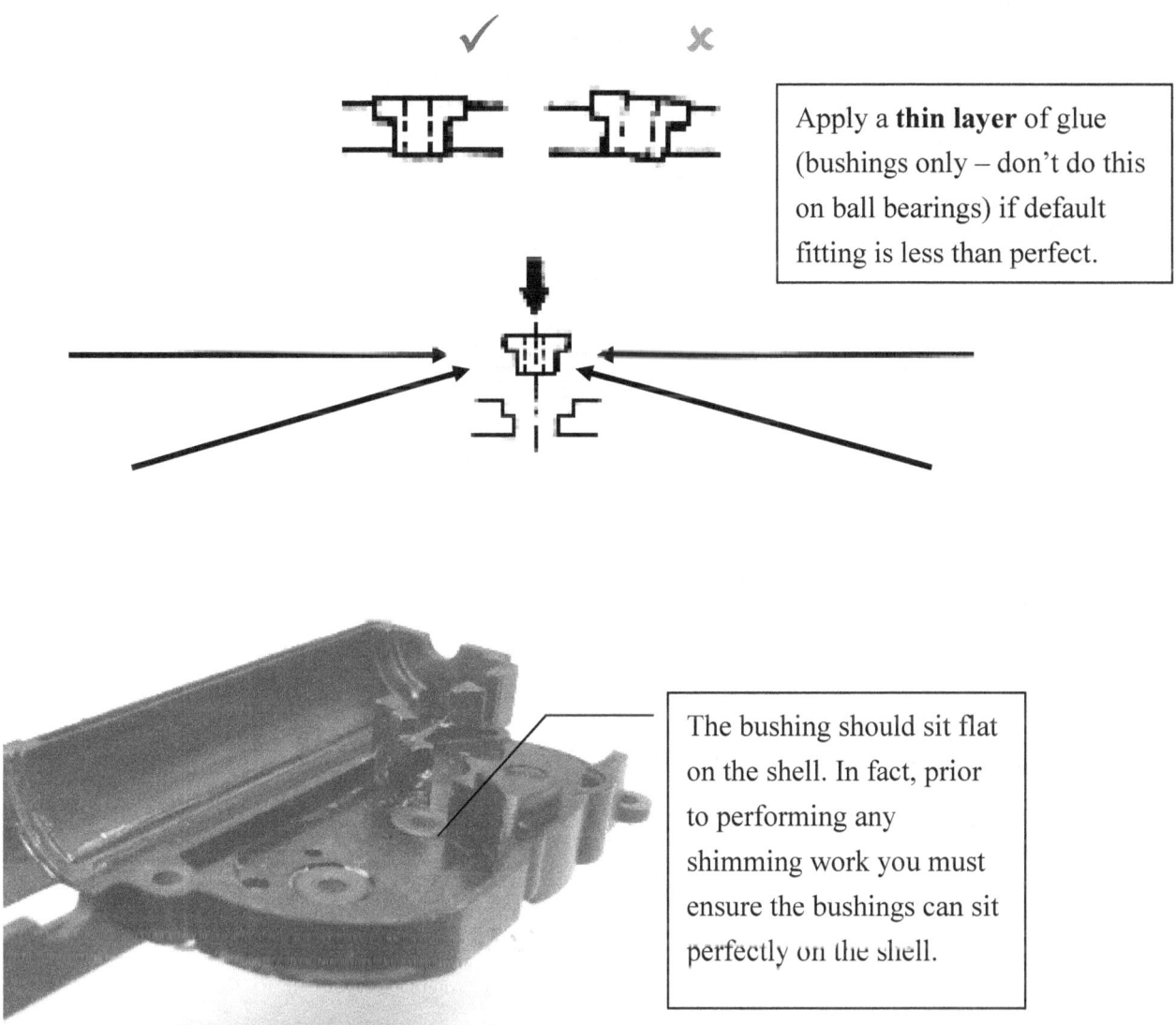

Apply a **thin layer** of glue (bushings only – don't do this on ball bearings) if default fitting is less than perfect.

The bushing should sit flat on the shell. In fact, prior to performing any shimming work you must ensure the bushings can sit perfectly on the shell.

Ultra FPS Technote

When you dremel the mechbox shell it is recommended that

you use those small scale portable type dremel tools. In fact, even a battery operated one (like the one shown below) is good. You don't want something too powerful as over-dremel would become too easy.

Get yourself a pack of 3mm shank grinding stone set. Start with the type shown on the left below and then move to the right one.

 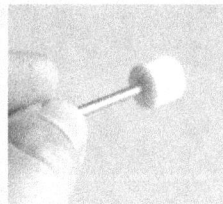

Do it VERY slowly, little by little. It should take you less than a few seconds to change a 6mm hole to a 7mm/8mm one. Remember, a tight fit is a MUST here.

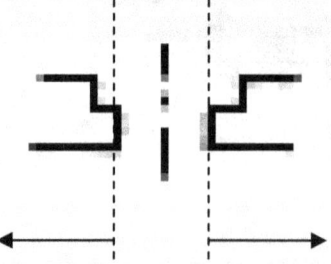

Apart from a dremel, you may accomplish the same through the use of other kinds of tools, such as drills, steel cutter and other hand tools...

If you are not sure what to use for this purpose, take the mechbox shell to your local hardware store and show the salespersons what needs to be accomplished. They will give you the best advice. Using professional tools such as milling machine and pressure pump is out of the scope of this text, but we will deal with those in our Advanced AEG Upgrade title.

Warning: Dremelling on the shell can be a high risk activity for beginners. Don't do it unless you have spare shell handy for you to practice with.

Buying third party replacement shell may come to the rescue if the original shell was screwed up. However, not all shells are created equal. Some small variances here and there can lead to problems everywhere.

Step 2: Shimming

Start by removing everything on the mechbox shell except for the bushings. By removing everything on the mechbox shell you can focus solely on the insertion and shimming of the 3 gears. An empty shell gives you a clearer view of the gears in action.

The reason why you need to carefully shim is that different makes of bushings and gears all have small variations in "thickness". Some gears may require a 0.2mm shim washer on one side while some others may require totally no washer on the top at all. The series of photos below are self explanatory – they show the suggested sequence of gear shimming. Just remember, the goal is to ensure the gears can interact smoothly without scratching each others. Also, nothing should scratch the items on the shell.

First you insert the spur gear (the middle gear), then the anti reversal. No scratching allowed here!

Then you insert the bevel gear and the sector gear. Again, no scratching with the spur gear allowed!

Double confirm that the gears are not scratching on each others.

Look at the setting from the side at different angles (as shown above). This way you can tell whether the teeth of the gears are meshing with the others properly.

Keep this in mind: when you shim the bevel gear pay attention to how many washers (and how thick these washers are) you put at its bottom. Remember, the motor pinion has to get in touch with it in order to turn the other gears. We cannot adjust the vertical alignment of the motor pinion, therefore if the bevel gear is shimmed too high the contact with the pinion may not be close enough to prevent stripping.

NOTE: **FYI:** when adjusting the position of the motor you should tighten it and test fire the gun bit by bit until the mechanism "sounds right". If the pinion cannot maintain close enough contact with the bevel gear (a setting which is too loose) a very strange noise will be generated when firing. On the other hand, if the setting is too tight the motor will stop working quickly (due to over heating).

When shimming has been initially completed you need to determine if any of the gears have been over-shimmed. Remember, when the gears run for a long time heat can make them "inflated". Therefore, at the time of shimming you need to leave room for accommodating such "inflation". Tight shimming can produce unexpected "runtime" problems.

You need to close up the shell tightly and manipulate the gears. Room for play should be

AirsoftPRESS (Hong Kong).

minimal but not totally tight (do leave a little room for play as things may expand a bit when heated up while running).

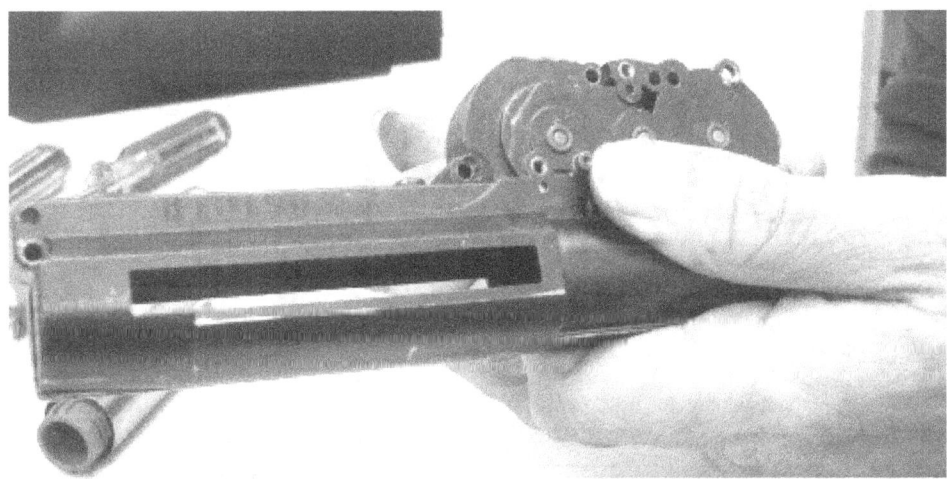

If you have difficulties putting the two halves of the mechbox back together, chance is that you have put too many shims somewhere. One extra needless shim on either side of any one

gear will usually make it impossible for the two halves to perfectly close.

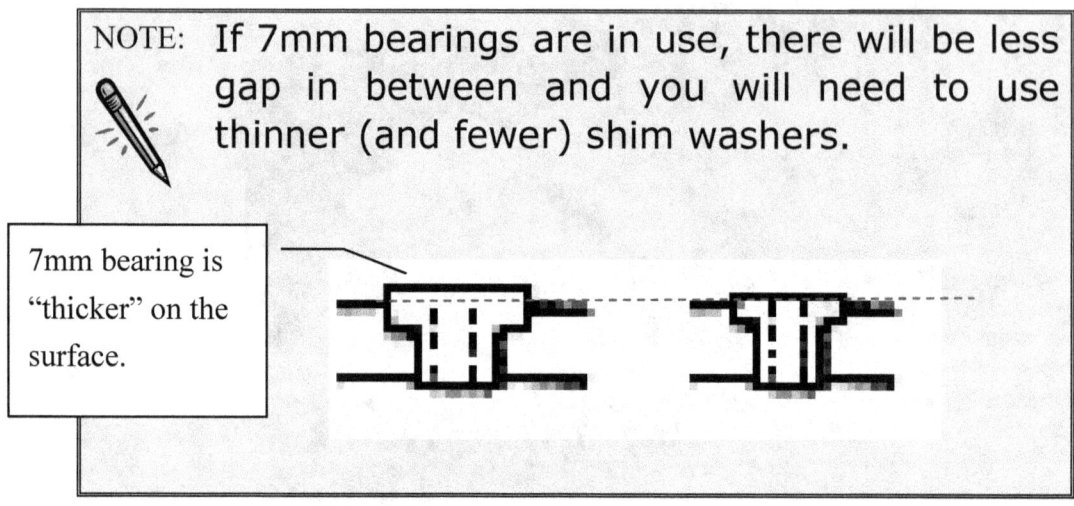

> **NOTE:** If 7mm bearings are in use, there will be less gap in between and you will need to use thinner (and fewer) shim washers.

7mm bearing is "thicker" on the surface.

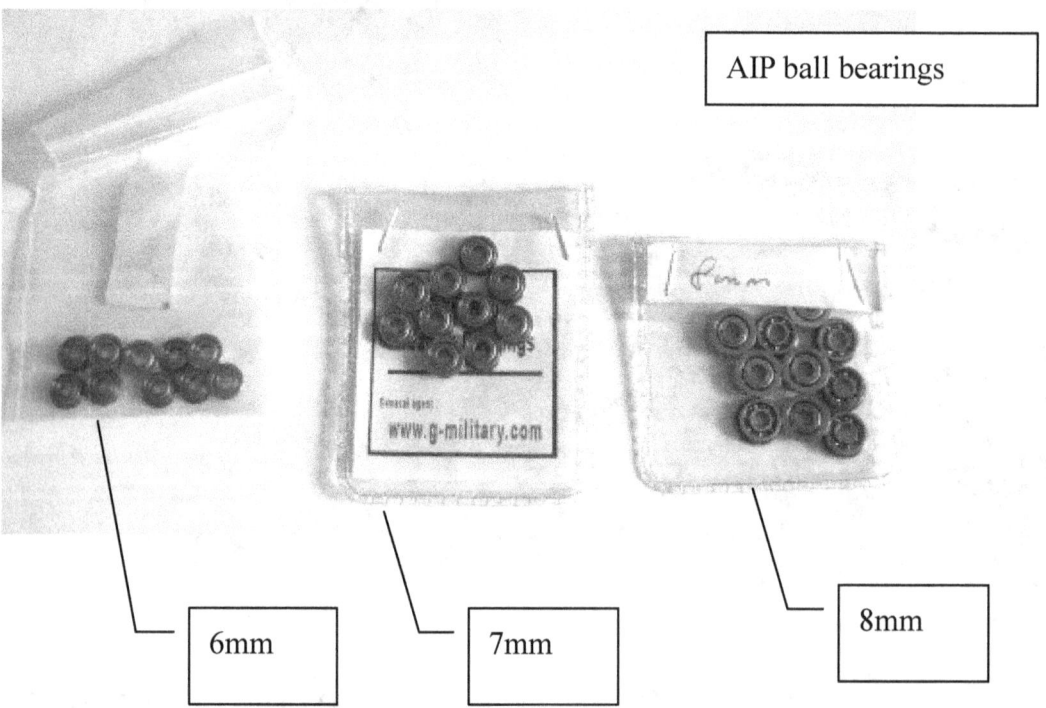

AIP ball bearings

6mm

7mm

8mm

It is quite common for people to manually enlarge the place

holders for the bushings so larger bearings can be installed. Generally speaking, 8mm ball bearings are more reliable if you have an upgrade spring installed.

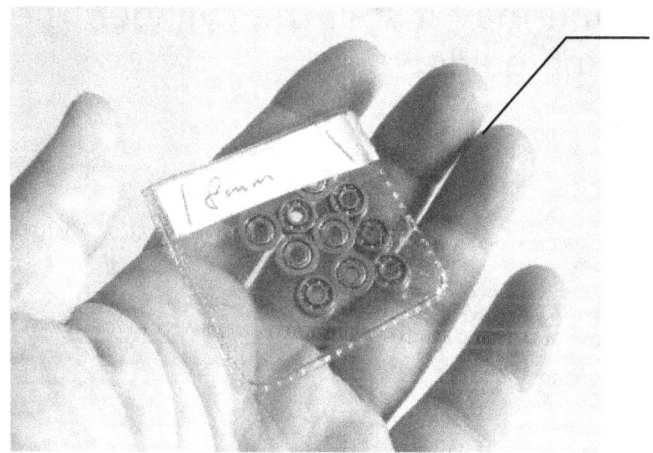

The AIP 8mm ball bearings are of pretty good quality.

Hold the 2 halves together tightly and see if the gears still have room to "play" from both sides of the shell. Use a pencil to manipulate the gear shafts. If they are okay initially, then tighten up the mechbox with screws and retry. If they can spin smoothly without much "play", write down the current setting on a piece of paper, remove the gears and washers and move on to the next step.

Do note that there is a reason to NOT shim too tight. A running mechbox has heat inside. Metallic parts when heated up may expand in size. You need to have room for accommodating such size increase.

Step 3: Installing the rest of the mechbox

Once the gears are in place you may install the cylinder, the piston, the spring and the spring guide:

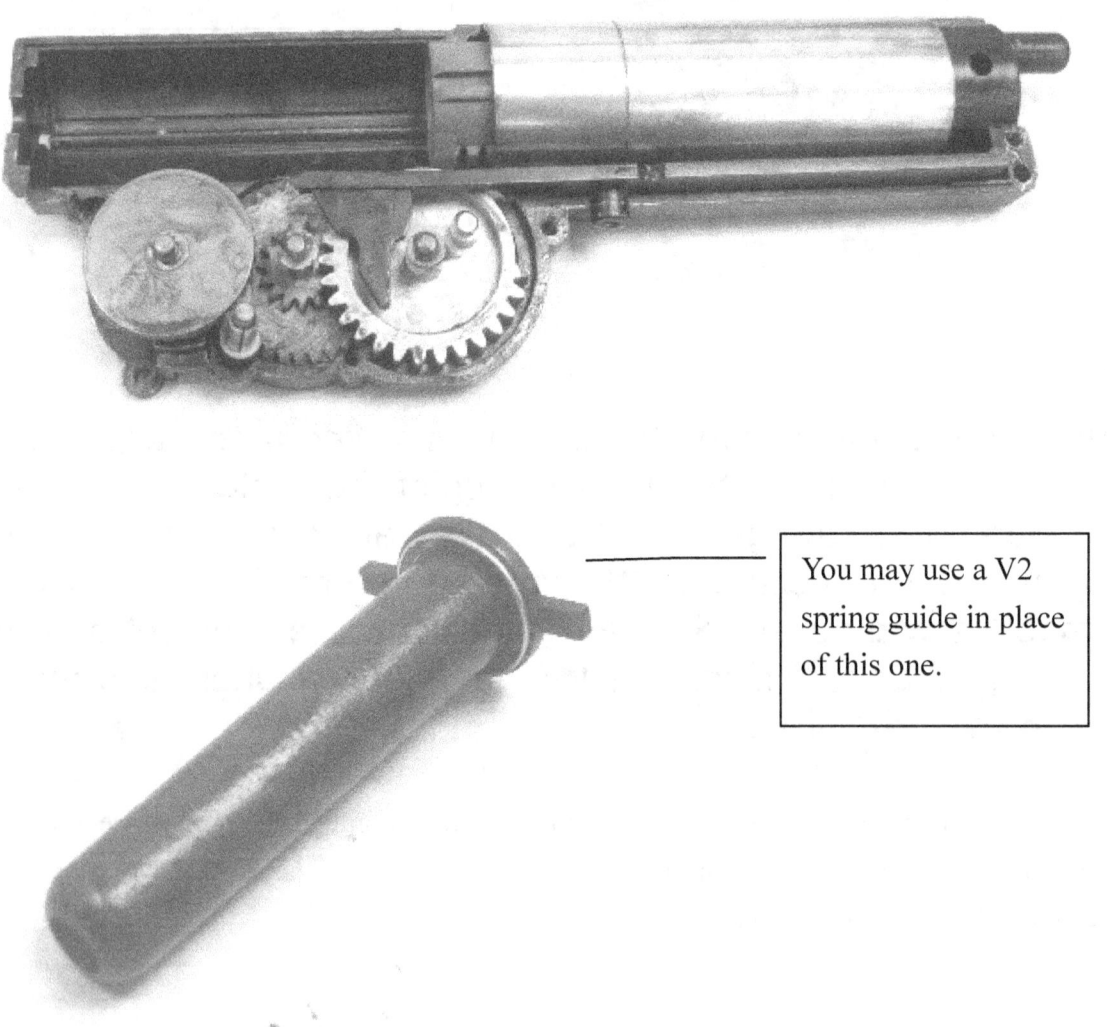

You may use a V2 spring guide in place of this one.

Prior to putting the mechbox back into the gun, you want to test fire it first.

Step 4: Testing

After closing up the mechbox, perform several tests manually to ensure that the mechbox has been properly assembled without glitches:

1, Use your finger to push the air nozzle inward. Can it move easily? And does it rebound by itself afterward? If it does not, that means something is wrong with the tappet plate – may be a mismatch of tappet plate and mechbox shell, a missing tappet plate spring, or that something has gotten in the way of the tappet plate. You will need to open up the mechbox again to find out what has gone wrong.

2, Can you feel a smooth trigger squeeze when manipulating the switch plate? If not, check the springs for use with these parts and make sure they have not been over-tightened.

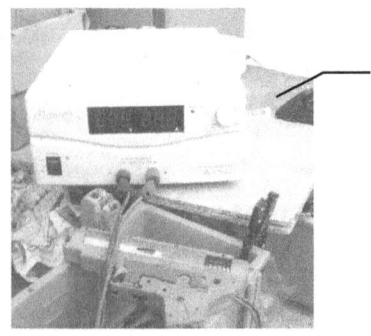

When you have a hard spring (such as a M110 or M120) installed, you need plenty of battery power to perform testing without locking up the mechbox. Professional tester use power supply device rather than battery for test shooting.

Step 5: Further troubleshooting

If after the upgrade your AEG shoots like this:

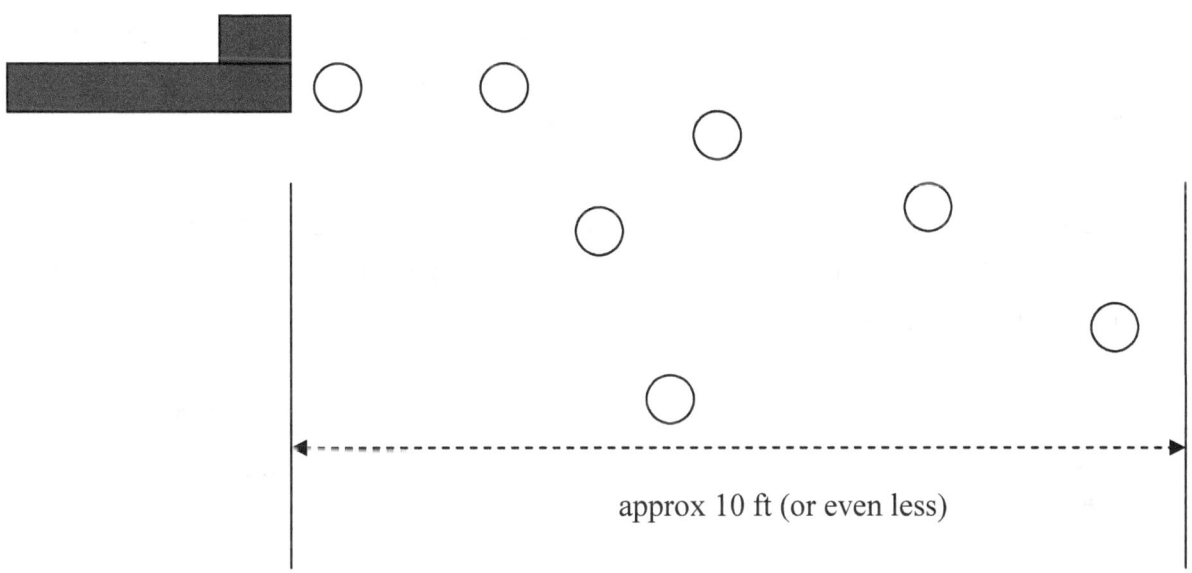

approx 10 ft (or even less)

<u>There are four major possibilities:</u>
i, damaged hopup - the rubber is dead, or the tiny spring which sustains the bucking is damaged. Practically speaking this seldom happens UNLESS you are using a China-made hopup rubber. There are many different grades of rubber and those China-made ones are just no good.

ii, grease is everywhere - at the tip of the air nozzle, around the hopup rubber, inside the barrel ...etc. **Highly likely after an upgrade job.** This is very easy to fix - just use the

AirsoftPRESS (Hong Kong).

cleaning rod.

iii, extremely poor alignment between the air nozzle and the hopup chamber - this can happen if the mechbox has not been seated properly on the receiver (some wires get in the way). This has happened with many AK implementations. You just have to rewire carefully and reseat the mechbox.

iv, the air nozzle is too short. Replace the nozzle and retry.

If you experience instant gun lockup:

First use a high power 9.6V or 10.8V pack to "unlock" it. If it does not work and you see smoke coming out of the wires, something is wrong with the gears. Possibilities include:

i, a loose sector gear delayer. Some sector gears just don't work with the delayer, gluing it is of no use, so your only option is to remove it.

ii, over shimming at the bevel gear.

iii, you left a screw somewhere inside the mechbox.

iv, a tappet plate that does not fit well with the shell. You either sand the sides of the tappet plate to make it fit, or replace it with one from another manufacturer.

If you experience gradual gun lockup, possibilities may include:

i, a loose sector gear delayer. Remove it.

ii, over shimming somewhere (or everywhere).

iii, poor quality / aging wires and/or connector plugs (not likely the case if your gun is pretty new).

iv, a spring that has not been properly grounded on either end (or on both ends). Refer to the picture below, the tip of the spring can prevent the spring itself from spinning, which could produce unnecessary tension on the gears. Check and make sure both ends of the spring are grounded (they should be flat), and that there is at least one metal ring spacer on the stock plastic spring guide to allow smoother spinning (apply a little bit of grease on both ends of the spacer too).

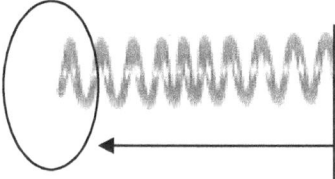

You may ground it yourself. Just clamp the tip and the last coil together and then heat it up with a lighter for 15 seconds or so.

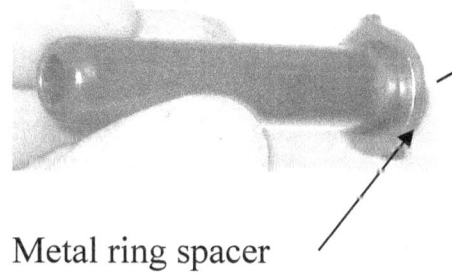

You don't need to use any spacer if you have a ball bearing spring guide.

Metal ring spacer

At the inside of the piston, if there is a spacer then make sure the surface of the spacer is completely flat. If you are using the TM type spacer (see the picture below), file the surface to

AirsoftPRESS (Hong Kong).   All rights reserved.

remove any unnecessary debris on it.

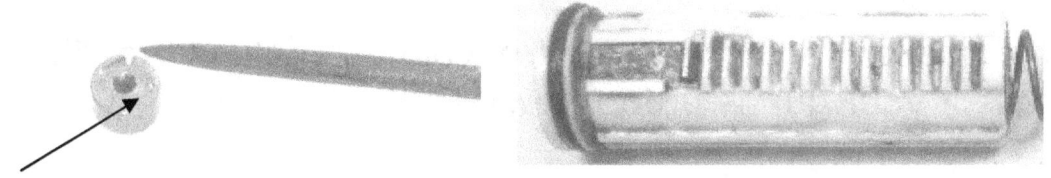

If the first 2 teeth of the piston break quickly:

i, your battery is way too powerful for the setup

ii, your piston is too brisky (this often happens with the China-made pistons).

Keep in mind, you won't fry your mechbox by using high power battery. You will only fry your mechbox ("burn" the switch plate) if something gets stuck inside.

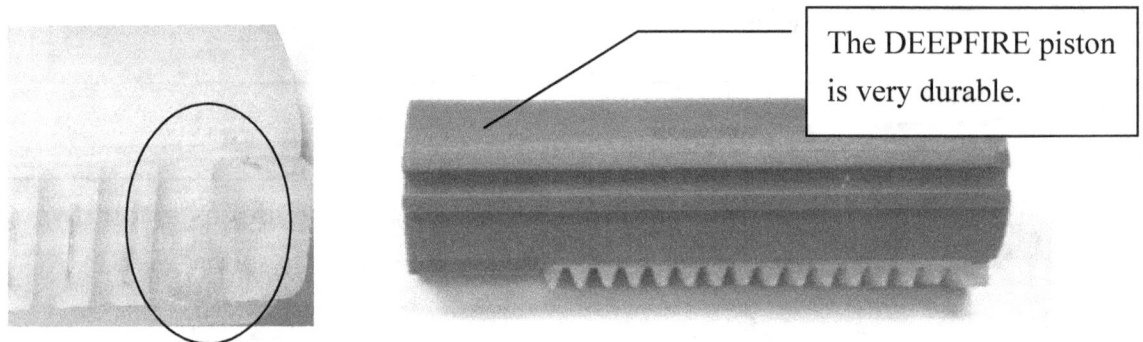

The DEEPFIRE piston is very durable.

What if you have no choice other than the small 1100mah cells?

Believe me, with these small cells even a 12V pack is very marginal for a PDI 170% (voltage will drop very quickly and the battery will get heated up like hell). This is especially true when your cells are from an unknown brand – many China-made battery clones are of very poor quality and would give you a way-faster-than-expected voltage drop.

The 1400 cells are slightly better but still, due to the small size heat becomes a big issue and your battery pack won't last long for even a half day game. Solution? A dirt cheap solution (assuming you already have the right air seal) is to do ALL of the following:

i, get a piece of stock spring from a WELLs/CYMA AEG. This stock spring is softer than a PDI spring, but is still stronger than a stock TM spring when configured with proper compression (through the use of spacers). The spring is easier to pull and is easier for the entire power system, especially when your battery is not in peak condition.

ii, use the stock TM piston, retain the piston spacer but make sure it has a totally flat surface facing the spring. Put 2 pieces of metal ring spacer (each about 1 to 1.5mm thick) on top of it.

iii, fit 5 to 6 pieces of metal ring spacer (each about 1 to 1.5mm thick) into the spring guide

Install the above into your mechbox and the gun will be able to produce the expected power without sacrificing reliability.

Special Topic: Maximum Air Seal

The importance of air sealing and air tightness

The goal of proper air sealing is to fill every gap along the air flow path to minimize air leakage and create an efficient environment for propelling BBs out of the barrel. To achieve this goal you need to first test for air tightness. Based on our experience higher end products from Marui, CA, ICS and the like are usually pretty good at that, but you can't expect the same from those cheaper China-made AEGs. Air sealing shall therefore become your top priority in gun upgrade and tuning.

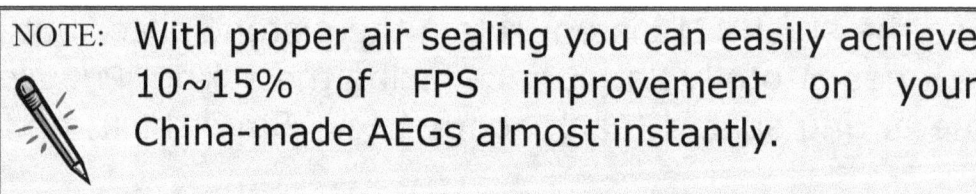

> NOTE: With proper air sealing you can easily achieve 10~15% of FPS improvement on your China-made AEGs almost instantly.

Air tightness is critical in the following two areas of the air flow path:

- Cylinder set
- Barrel and Hopup

Air tightness at the cylinder set

To test for air tightness at the cylinder set, block the cylinder head nozzle with a finger and then try to push the piston.

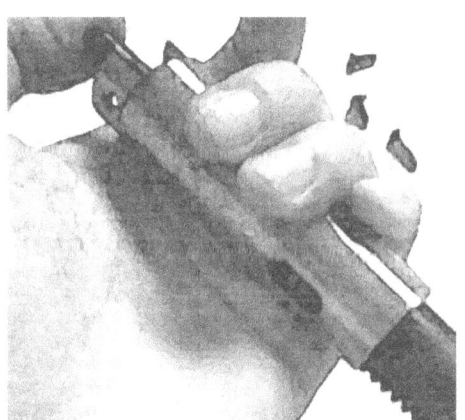
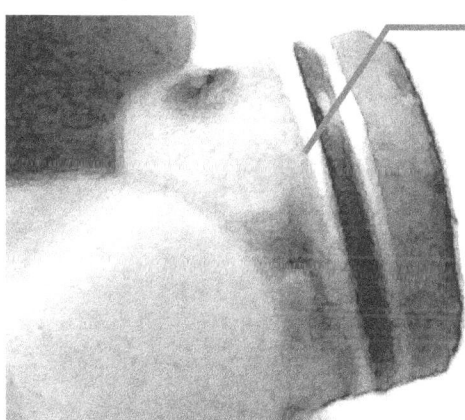

Pipe wrap behind the O ring

If you experience strong resistance, air tightness is good. If not, two possibilities exist:

I, Air leaks through the gap between the cylinder head and the cylinder. To prevent leakage of this sort, you use a thin layer of pipe wrap to wrap around the side of the cylinder head before putting in the cylinder head O ring:

II, Air leaks through the piston O ring (this is the most common cause of performance problem among the China-made AEGs). A properly sized replacement can usually fix the problem.

Air tightness at the barrel & chamber set

To test for air tightness at the barrel and the hopup chamber, you block the chamber entrance and the BB loading nozzle with your fingers and then try to blow air into the barrel:

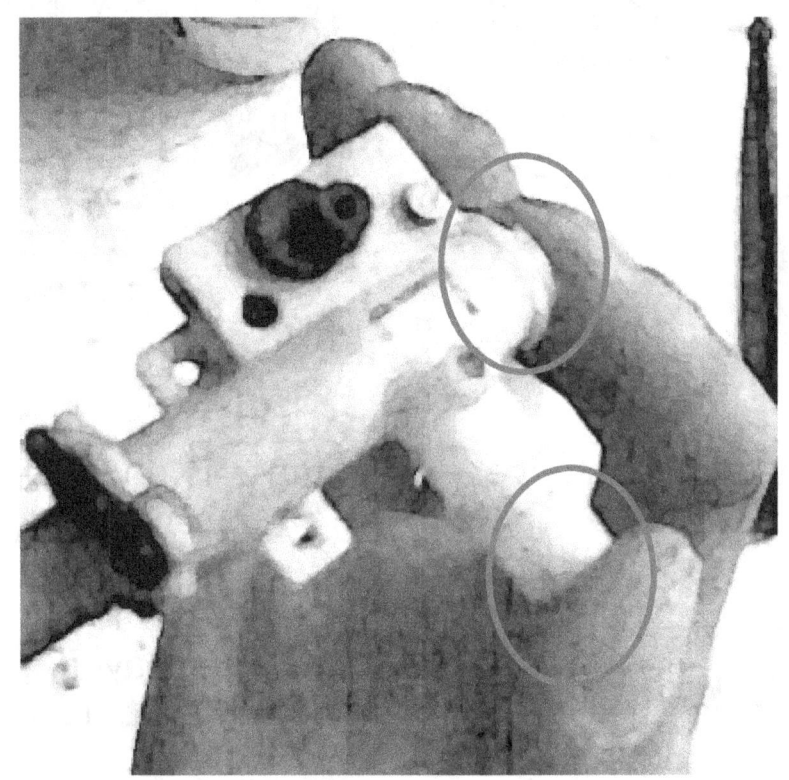

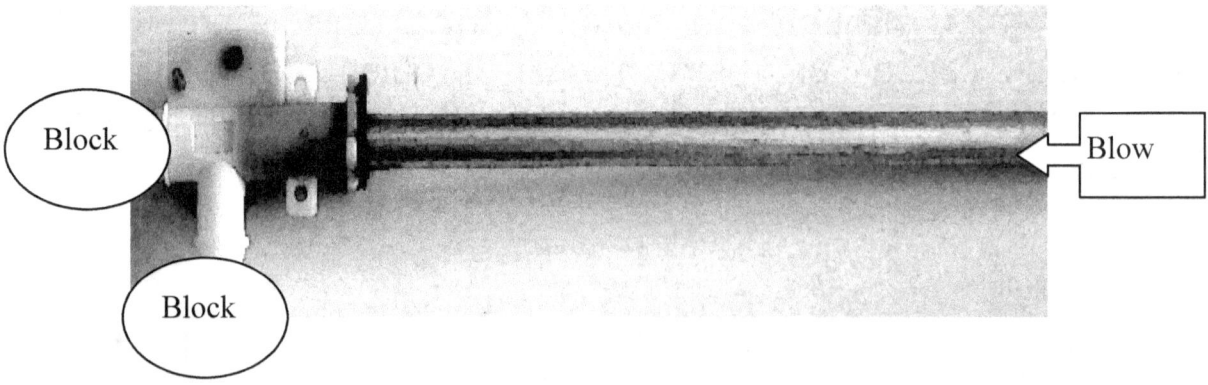

Block

Block

Blow

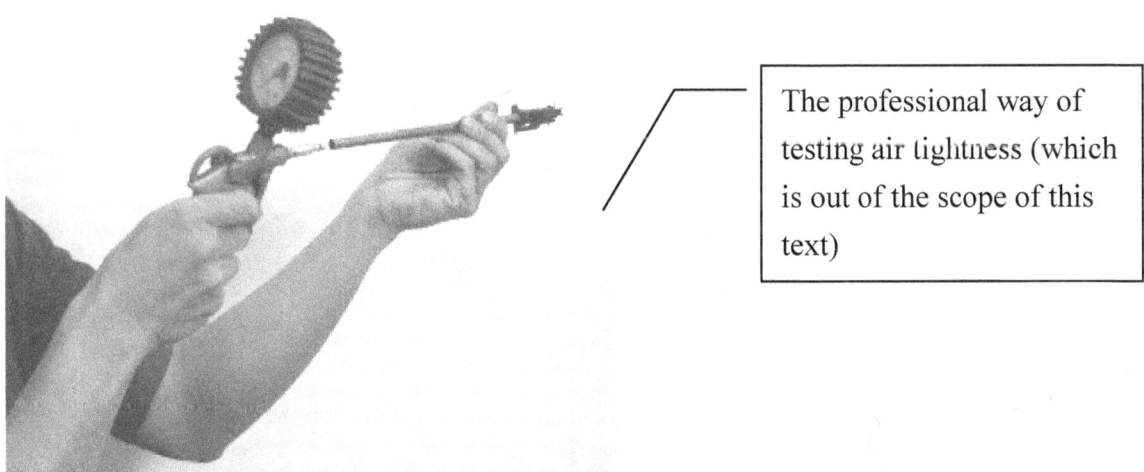

The professional way of testing air tightness (which is out of the scope of this text)

If it is real tough for air to go into the barrel, air tightness is good. If not, you may want to do the following:

I, Before attaching the hopup rubber to the barrel, apply a thin layer of grease around the barrel (but make sure the grease doesn't block the barrel opening) to fill any gap between the inner side of the hopup rubber and the outer surface of the barrel:

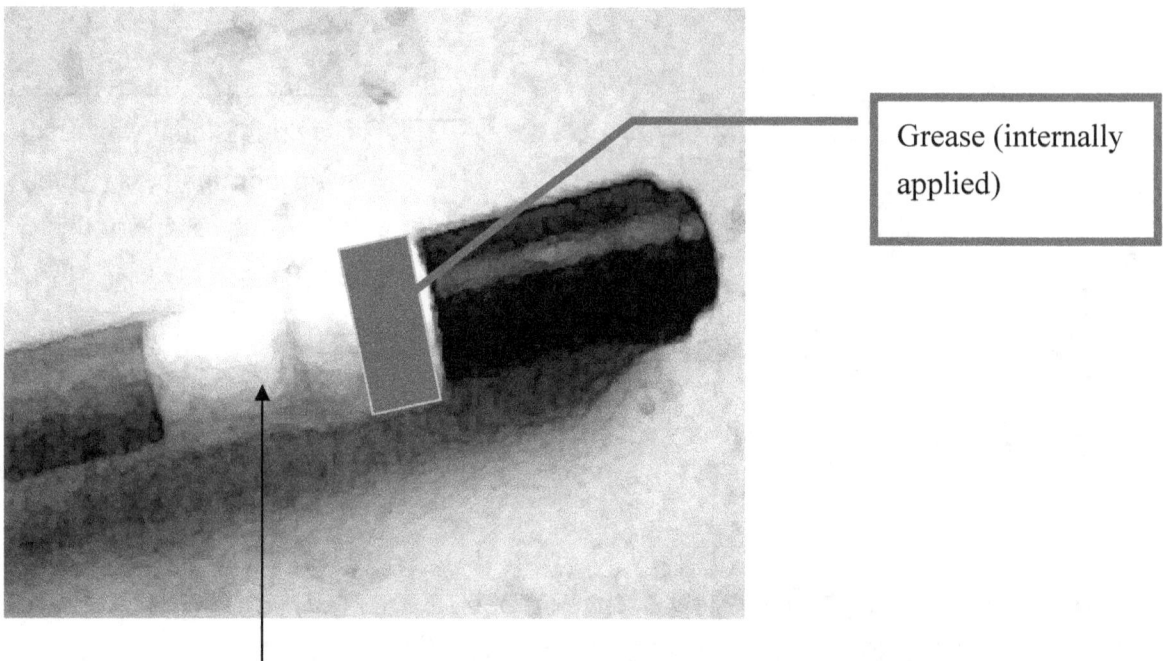

Grease (internally applied)

II, Now use pipe wrap to wrap around the barrel and the hopup rubber to further prevent air leakage.

III, You may also want to improve air tightness at the air nozzle. The easiest thing to do with your stock nozzle is to apply a very thin layer of grease around the inner side of the nozzle (and most importantly - without blocking the opening):

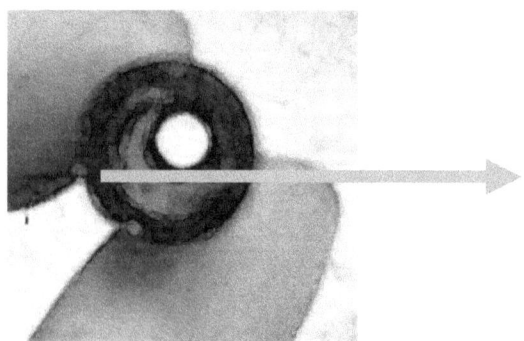

Warning: Don't do silly things like inserting a small O ring into the stock O-ring-less nozzle. An air seal nozzle isn't created this way. Any mismatch in size/shape here can jam the mechbox and screw up the internals.

High quality nozzles always have O rings "built-in"! Refer to the picture below, the left one has built-in O ring. The right one does not.

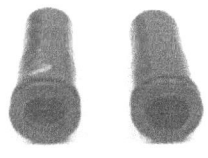

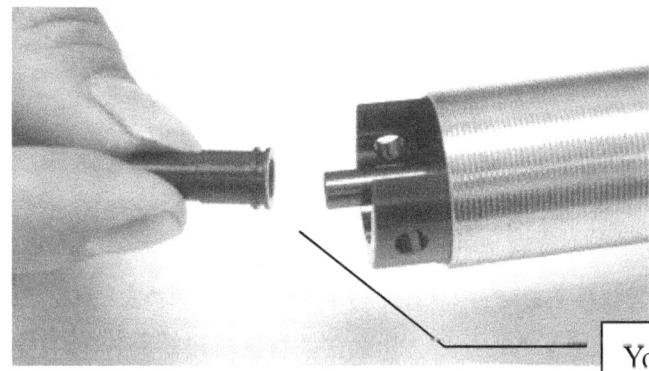

You must be very precise here – air leakage can be very serious at the nozzle.

Special technote on piston head modification

Upgraded piston heads usually have holes on the front face of the head. When the piston is moving, air goes through these holes and forces the surrounding O-ring outwards a tiny bit to create a seal for preventing air leakage (this is how the one-way mechanism works).

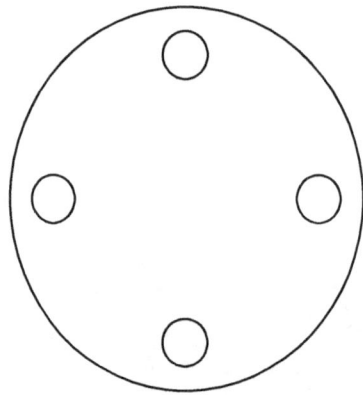

The WELLS piston head has the holes drilled for you already so you don't need to modify the piston head any further (except to replace the O ring when necessary). More holes on it would not be necessary. On the other hand, make sure you don't over-apply grease or these holes will be accidentally blocked.

The Pro Ology high end piston head has a different design – it has a proprietary one-way design where the holes are actually hidden behind the O ring.

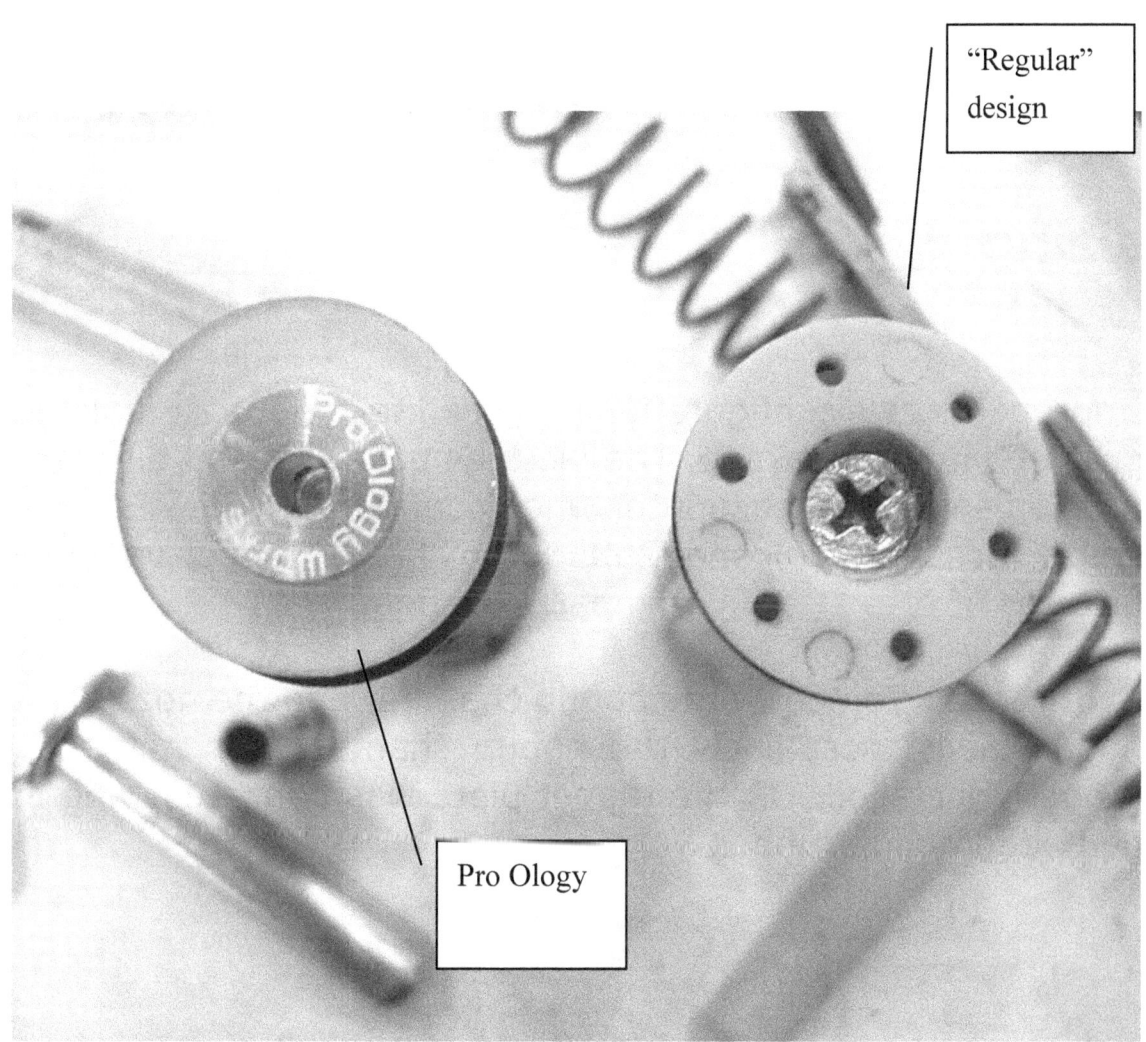

"Regular" design

Pro Ology

Special technote on air nozzle shape and sizing

Improper nozzle configuration may lead to serious air leakage or even hopup breakage. The pictures below show 3 different problems you may encounter when configuring your AEG (full credit goes to zhigangd2005 who posts these valuable pictures at http://forum.combat2000shop.com/:

Problem 1: Somehow the hopup chamber and the nozzle are not leveled (most likely the hopup chamber has not been properly installed, OR the tappet plate was bent, resulting in damaged hopup rubber.

Notice the angle of the nozzle

Problem 2: The nozzle is too short, resulting in air leakage.

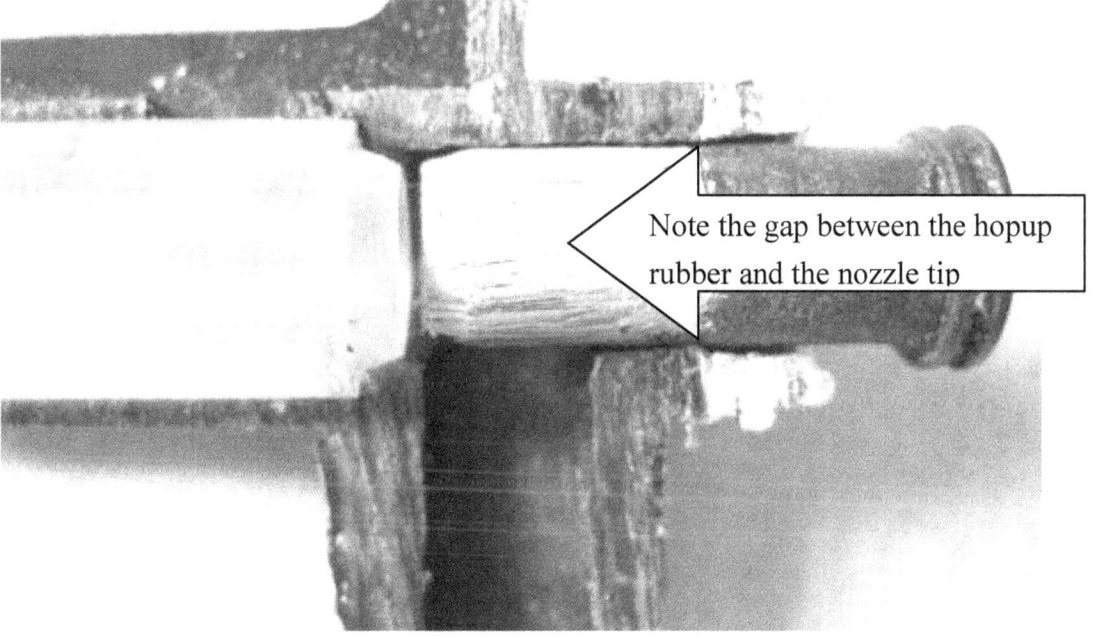

Note the gap between the hopup rubber and the nozzle tip

Problem 3: The air nozzle is too long, and the nozzle tip has a shape that does not fit well with the opening of the hopup rubber.

Mismatch between the nozzle tip and the hopup rubber

The picture below shows 3 different air nozzles produced by 3 different Chinese manufacturers for the same gun model. They are all different, even though their hopup chambers are similarly structured.

Special technote on motor and gears break-in

You want to first break-in the brush/commutator interface so that the brushes can conform better to the shape of the commutator. To do so is easy – just run the motor on 4 cells (1.2V on each cell) for several minutes until the full brush face is conformed to the commutator.

Breaking in bushings is necessary if the motor bushings are too tight. To do a quick check, just spin the motor (by hand) with the brushes removed and fell the resistance to turning (you may want to have some other motors here for comparison purpose).

To perform bushing break-in, just put a little valve of grinding compound into the bushing and spin the motor until you feel a reduction in resistance.

You may also want to break-in the mechbox gears for smoother operation. To break in these gears, setup the mechbox in such a way that only the bushings and the gears are in place. No anti-reversal, no piston and no other parts. Then, close the mechbox shell tightly and hold the motor to the bevel gear. Have the motor driven by 6V or 7.2V power (with excellent shimming your motor should be able to drive the gears at 6V), and let the motor pinion drives the gears. Of course, before you do this you want to ensure that all these gears (including the motor pinion) have been lubed. Let them spin like this for a couple of minutes and the gears break-in process is considered completed.

For the latest product releases, please visit:

http://www.airsoftpress.com

Please email your questions and comments to

editor@airsoftpress.com

Thank you.

RCPRESS publishes books on RC technology.

www.rcpress.com